D1189629

HIGH MODULUS FIBER COMPOSITES IN GROUND TRANSPORTATION AND HIGH VOLUME APPLICATIONS

A symposium
sponsored by
ASTM Committee D-30
on High Modulus Fibers
and Their Composites
Pittsburgh, PA, 7 Nov. 1983

ASTM SPECIAL TECHNICAL PUBLICATION 873
D. W. Wilson, University of Delaware,
editor

ASTM Publication Code Number (PCN)
04-873000-33

1916 Race Street, Philadelphia, PA 19103

Library of Congress Cataloging in Publication Data

High modulus fiber composites in ground transportation and high volume applications.

(ASTM special technical publication; 873)
"ASTM publication code number (PCN) 04-873000-33."
Includes bibliographies and index.
1. Fibrous composites—Congresses. I. Wilson, Dale W. II. ASTM Committee D-30 on High Modulus Fibers and Their Composites. III. Series.
TA418.9.C6H54 1985 620.1'18 85-6030
ISBN 0-8031-0415-4

NOTE

The Society is not responsible, as a body,
for the statements and opinions
advanced in this publication.

Printed in Baltimore, MD
June 1985

Foreword

The symposium on High Modulus Fiber Composites in Ground Transportation and High Volume Applications was held in Pittsburgh, Pennsylvania on 7 Nov. 1983. ASTM Committee D-30 sponsored the symposium. D. W. Wilson, University of Delaware, Center for Composite Materials, presided as symposium chairman. D. L. Denton, R. E. Evans, C. D. Shirrell, and S. S. Wang presided as session chairmen.

Related ASTM Publications

Long-Term Behavior of Composites, STP 813 (1983), 04-813000-33

Analysis of the Test Methods for High Modulus Fibers and Composites, STP 521 (1973), 04-521000-33

Effects of Defects in Composite Materials, STP 836 (1984), 04-836000-33

A Note of Appreciation to Reviewers

The quality of the papers that appear in this publication reflects not only the obvious efforts of the authors but also the unheralded, though essential, work of the reviewers. On behalf of ASTM we acknowledge with appreciation their dedication to high professional standards and their sacrifice of time and effort.

ASTM Committee on Publications

ASTM Editorial Staff

Contents

Introduction

The conference on High Modulus Fiber Composites in Ground Transportation and High Volume Applications focused on a very important issue in composites technology, the relationship between processing, material microstructure, and the resulting material properties. Material formulations and processing techniques employed in the high volume manufacturing of composite components produce complex microstructure, but without control. The design of material microstructure, especially fiber orientation state, forms the basis for effective composites utilization, and ineffective control of microstructural parameters increases property variability and forces poorly optimized designs. Ultimately, high volume processes need to be developed which allow design and control of material properties during processing.

The first steps toward this goal are to understand the relationship between processing parameters and material microstructure and to be able to quantitatively measure and describe microstructural features which determine the macroscopic material properties. The relevant microstructure parameters which control composite properties are the constituent properties, volume fractions of constituents, interfacial adhesion between the phases, fiber orientation state, fiber aspect ratio, void content, state of resin cure (thermosets), and resin crystallinity (thermoplastics). In two-phase, continuous fiber aerospace composites aspect ratio and fiber orientation are highly controlled. Composites for high volume applications are often discontinuous fiber, three-phase systems in which fiber aspect ratio and orientation state are not controlled. The control of these two parameters and understanding the effects of their distribution on properties is a central problem in the effective use and fabrication of high volume composite materials.

The intent of the conference was to provide a forum for discussing the macroscopic behavior of typical high volume composite material systems, materials characterization methods, and the relationship between processing and material properties. The knowledge developed in these areas will form the foundation for the ultimate objective, design and control of composite properties.

The papers published in this volume address these primary topics along with more advanced property characteristics such as environmental sensitivity, fatigue behavior, and viscoelastic response. The identification and systematic research of the complex topics addressed in this volume is embryonic. It is hoped that the reported research findings will serve to stimulate new, broader based research activities in this important area of composites technology.

D. W. Wilson

Associate scientist and assistant director, University of Delaware, Center for Composite Materials, Newark, Delaware; symposium chairman and editor.

C. David Shirrell [1]

The Influence of Microstructural Variability upon the Scatter in Mechanical Properties of R25 Sheet Molding Compound

REFERENCE: Shirrell, C. D., **"The Influence of Microstructural Variability upon the Scatter in Mechanical Properties of R25 Sheet Molding Compound,"** *High Modulus Fiber Composites in Ground Transportation and High Volume Applications, ASTM STP 873,* D. W. Wilson, Ed., American Society for Testing and Materials, Philadelphia, 1985, pp. 3–22.

ABSTRACT: The large scatter in the static strengths of R25 sheet molding compounds was observed to be related to the variability in the complex microstructure of this material. This microstructure consists of a subsurface veil of individual glass fibers and an internal core of intact fiber glass bundles.

KEY WORDS: sheet molding compounds, discontinuous composites, composites variability, composite microstructure

Short, chopped glass fiber reinforced sheet molding compound (SMC) is moving into its second generation of mass automotive applications. Previously, the use of this material has been limited (in large volumes) to nonstructural, appearance components such as front end header panels. In the newer second generation applications, SMC will be used in lightly stressed structural components (such as tailgates and body panels) [*1*,*2*]. Future design developments using this material in highly stressed major automotive components may lead to a third generation of SMC applications.

These more demanding second and third generation structural applications of SMC are placing increased emphasis on the elimination of the large mechanical property variability found in this material. Previous work in the composites literature has documented that this material has scatter in its mechanical properties as large as 40% even when such obvious material defects as weld lines and gross anisotropic glass fiber orientation have been eliminated [*3*,*4*]. While this

[1]Staff research engineer, Polymers Department, General Motors Research Laboratories, General Motors Technical Center, Warren, MI 48090–9055.

scatter in mechanical properties can be accurately described by empirically derived Weibull statistical parameters, simple probabilistic flaw site models do not correctly predict the observed mechanical property variability in this material [4].

Thus, the nature of the flaw site(s) that leads to the large scatter in mechanical properties of SMC remains unknown. In an attempt to partially resolve this issue, the microstructures of three different R25 (25% by weight randomly oriented glass fibers) SMC materials were examined in detail, and this paper will discuss those results and their relevance to the scatter in mechanical properties of SMC.

Experimental Procedures

Material Fabrication

The details of the three R25 SMC compositions used in this study have been described previously [4]. Two of these materials utilized isophthalic polyesters as their resin matrix, while the third was formulated with a vinyl ester resin matrix. All of these SMC materials were molded in the form of flat plaques with dimensions 533 by 610 by 3.4 mm (21 by 24 by 0.13 in.). The molding procedure consisted of charging approximately 2000 g (4.4 lb) of the uncured SMC material into the center of a compression die preheated to 138°C (280°F) and applying a pressure of 5.5 MPa (800 psi). The SMC was held at these conditions for 120 s to complete the cure of the plaque.

Specimen Fabrication

These cured plaques of SMC materials were cut into ASTM Method for Tensile Properties of Plastics (D 638-82a) Type 1 tension specimens, ASTM Methods for Flexural Properties of Unreinforced and Reinforced Plastics and Electrical Insulating (D 790-81) flexure specimens, and 25.4 by 25.4 mm (1 by 1 in.) density and burn-out specimens on a water-cooled diamond saw using the procedures which have been previously described [4].

Experimental Procedures

Density, resin content, filler content, and fiber content of the SMC materials were determined from each of the panels used in this study. The results of these tests can be found elsewhere [4].

Mechanical Testing Procedures

Prior to mechanical testing, all tension and flexure specimens were weighed and then maintained at 63°C (145°F) and 0% relative humidity until they reached a constant weight. The specimens were then tested at room temperature (≈21°C) in either uniaxial tension, three-point flexure, or four-point flexure on an Instron Universal Test Machine using the procedures described in either ASTM Method for Tensile Properties of Plastics [Metric] (D 638M-81) or ASTM Methods for Flexural Properties of Unreinforced and Reinforced Plastics and Electrical Insulating (D 790-81) (Method I and II). The flexure specimens were tested using an L/d (span/depth) ratio of 16 to 1. Tension specimens which did not fail in the gage area (fracture site index numbers greater than 7.5 and less than 14.5) were deleted from the tensile data.

Results and Discussion

Statistical Aspects of SMC Strengths

The variability in mechanical properties of SMC is substantial. One of the polyester resin SMCs used in this study has coefficients of variation in mechanical properties which are as large as 18% (Table 1). As indicated in Fig. 1, the specimen orientation does not appear to significantly influence this mechanical property variability. (For brevity, only the data from one SMC material will be presented in this paper. Unless otherwise noted, the other SMC materials examined in this study exhibited similar results.) Thus, gross fiber anisotropy can be eliminated as a source of the mechanical property scatter in these plaques.

TABLE 1—*Strength (σ) and modulus (E) of R25 SMC.*[a]

Mechanical Property		Statistical Measure[b]		
		Average	Standard Deviation	Coefficient of Variation, %
			POLYESTER TYPE I	
Tensile	σ	48.8	9.1	18.6
	E	11.7	1.2	10.3
3-point flex	σ	131.9	19.4	14.7
	E	10.5	0.5	4.8
4-point flex	σ	116.9	21.1	18.0
	E	14.2	1.2	8.5
			POLYESTER TYPE II	
Tensile	σ	56.2	6.9	12.3
	E	14.1	1.5	10.6
3-point flex	σ	156.1	22.5	14.4
	E	13.9	0.7	5.0
4-point flex	σ	137.2	18.3	13.3
	E	15.5	0.9	5.8
			VINYL ESTER	
Tensile	σ	96.0	8.6	9.0
	E	13.7	1.2	8.8
3-point flex	σ	216.6	29.1	13.4
	E	14.3	0.9	6.3
4-point flex	σ	159.6	18.0	11.3
	E	16.4	0.6	3.7

[a]Strength in MPa and modulus in GPa.
[b]The number of replicates ranged from 22 to 39.

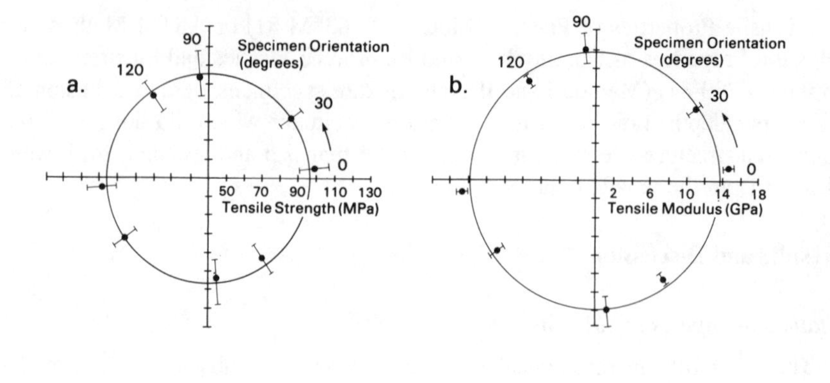

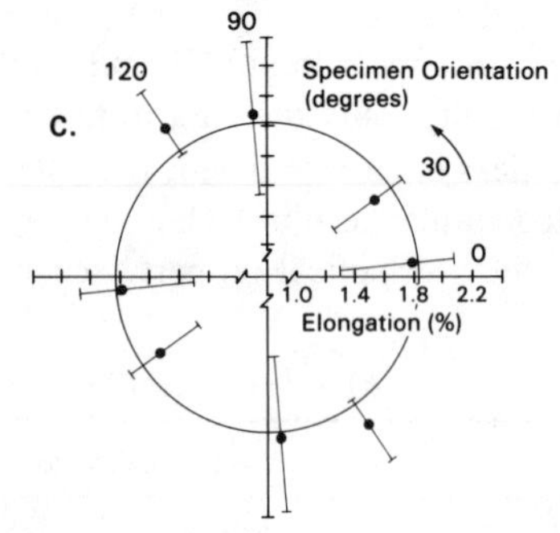

FIG. 1—*Polar diagram of specimen orientation versus tensile properties for the vinyl ester SMC:* (a) *strength,* (b) *modulus, and* (c) *elongation.*

In Fig. 1, the average value of six replicates per orientation is indicated by a dot. The scatter band represents the range of the data. For these polar plots, a deviation from a perfect circle, which has a radius equal to the average property value, represents a variation in mechanical properties with specimen orientation. The observed variability in mechanical properties of this material is also not due to gross variations in either void content, resin content, filler content, or fiber content of the SMC plaques used in this study [*4*].

Rather, the source of the mechanical strength variability in SMC materials appears to be randomly occurring internal material flaws. Previous work [*4*] has shown that Weibull statistics can provide a useful technique for quantifying and interpreting the nature of these material flaws in R25 SMCs. One of the Weibull strength distributions for the vinyl ester SMC examined in this study is given in Fig. 2. As discussed in Ref *4*, it is possible to calculate the Weibull strength distribution using theoretical flaw site models. Figures 3 and 4 illustrate that these theoretical flaw models, based on random flaws distributed in either the surface, volume, or edge of the SMC mechanical specimens, do not predict the experimentally observed results.

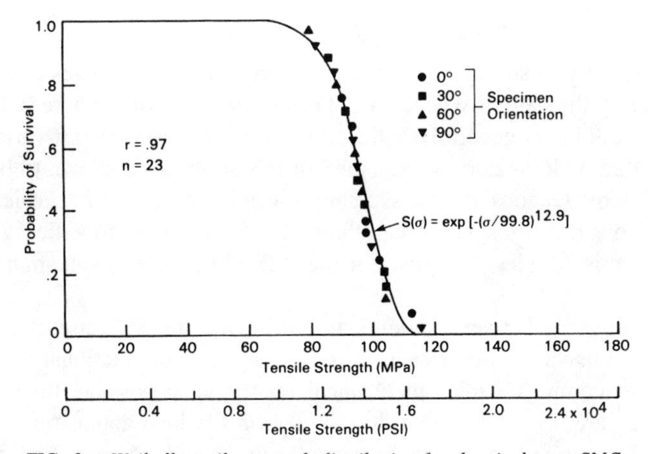

FIG. 2—*Weibull tensile strength distribution for the vinyl ester SMC.*

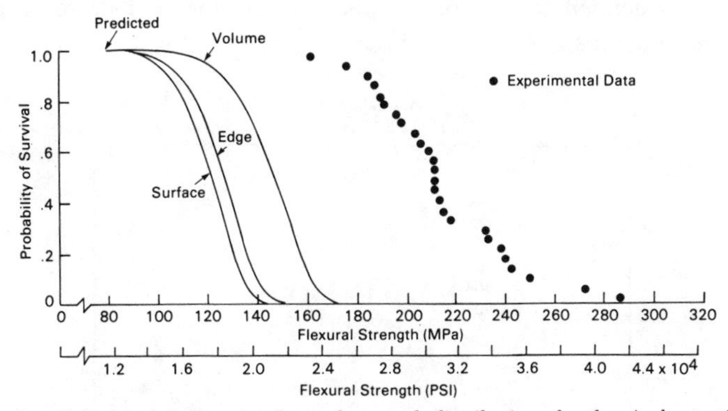

FIG. 3—*Calculated three-point flexural strength distributions for the vinyl ester SMC.*

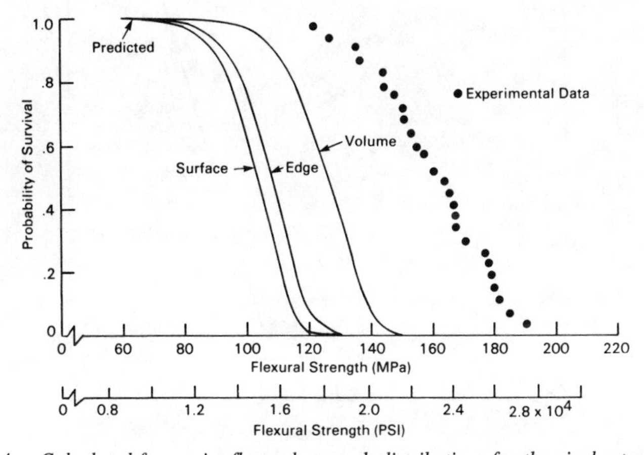

FIG. 4—*Calculated four-point flexural strength distributions for the vinyl ester SMC.*

Strength and Modulus Relationships for R25 SMC

The variability in strength of random discontinuous composites appears to be determined by the naturally occurring flaws of this material, whereas the scatter in elastic modulus is generally believed to be related to material variation [*5*]. Since the R25 SMC materials examined in this study were shown to be macroscopically homogeneous and transversely isotropic, the scatter in elastic modulus should be considerably smaller than the scatter in strength properties. As shown in Table 1, this is indeed the case for the R25 SMC materials examined in this study.

The relationship between moduli and strength can be a useful method to determine if this variability is caused, in part, by fiber orientational variations in SMC. From composite micromechanical theory, it is possible to predict that fibers which are oriented parallel to the edge in the load application section of three- and four-point flexure specimens will substantially increase both the specimen's strength and modulus [*6*]. Those fibers which are oriented transverse to the edge of the specimen in the load application section will decrease both the strength and modulus.

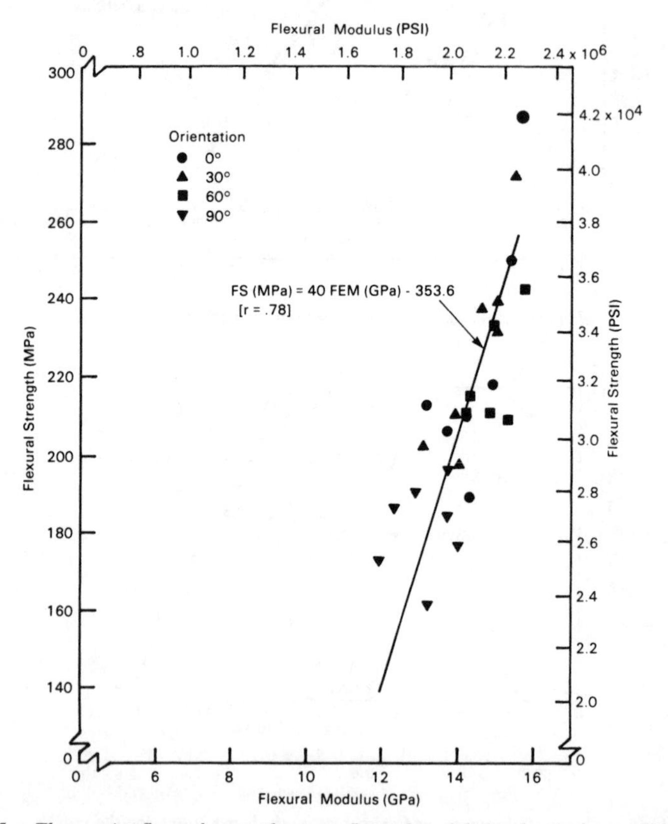

FIG. 5—*Three-point flexural strength versus flexural moduli for the vinyl ester SMC.*

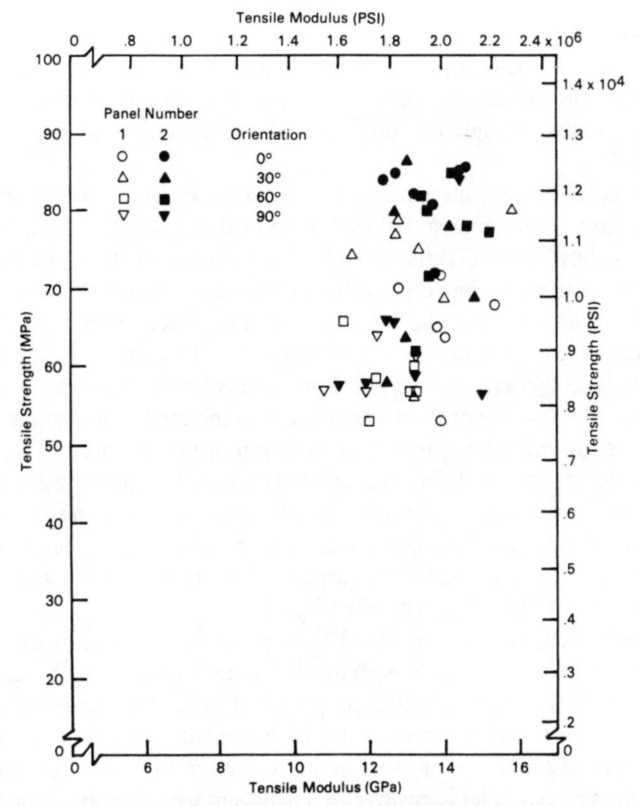

FIG. 6—*Tensile strength versus tensile moduli for a polyester SMC.*

Thus, the presence of any locally oriented glass fibers in SMC should result in a positive correlation between strength and modulus in three- and four-point flexure specimens (Fig. 5). There appears to be a statistically significant ($\alpha = 0.005$) linear relationship between flexural strength and flexural modulus in three of the six flexural data sets examined in this study.

However, there should be no relationship between the strength and modulus of tension specimens since the load in this type of specimen is distributed across a much larger testing section than in either type of the flexure specimens. In this larger testing section, the nature and number of the flaw sites determine the strength of the specimens, whereas the homogeneity of the fiber orientation and the fiber distribution across the testing section determine the specimen's modulus. Thus, as shown in Fig. 6, there is apparently no relationship between strength and modulus in R25 SMC tensile data.

Relationship Between SMC Microstructure and Mechanical Properties

The three-point flexure specimens tested in this study failed on the tension (that is, the lower) face of the specimen opposite to the loading nose of the test fixture.

This would be the anticipated failure mode since the maximum stress occurs in the outer fibers of the beam along its lower face. Thus, the fracture which caused these specimens to fail was initiated at the tensile surface of the specimen and grew diagonally through the thickness of the specimen as the failure progressed [*4*].

In four-point flexure, the maximum stress also occurred near the tension face of the specimen over an area defined by the load span (that is, distance between the two loading noses). Thus, the observed fracture of these specimens was initiated at a site approximately one third of the way through the thickness of the specimen, measured from the tension face. This fracture grew parallel to the neutral axis of the specimens via delamination as the failure progressed [*4*].The failure mode in the tension specimens is more complex than that in the flexure specimens. The fracture of most specimens was initiated at or near the surface of the specimen in the tension tests while crack propagation through the thickness of the specimens occurred via a repeating cycle of (*a*) vertical crack growth until a fiber bundle was struck, (*b*) interlaminar crack growth along the glass/resin interface until the crack reached the end of the bundle, and (*c*) continued vertical crack growth. The resulting failure surfaces of the tension specimens were characterized by their jagged appearance (Fig. 7).

Thus, the strength and modulus of three-point flexure specimens, four-point flexure specimens, and tension specimens are each determined by different regions in the SMC. The mechanical properties of SMC under three-point flexural loading are sensitive to the nature of any microstructural variation near the tensile surface of the specimen. However, in the four-point flexure specimens the mechanical properties are less sensitive to tensile surface microstructural variations and more sensitive to interior microstructural differences. The mechanical properties of SMC in tensile loading are sensitive to microstructural variations throughout the length and cross-sectional area of the specimen.

A large number of fractured three-point flexure specimens, four-point flexure specimens, and tension specimens were subjected to pyrolysis followed by treatment with hydrochloric acid (HCl) mist spray (for removal of the calcium carbonate [$CaCO_3$] filler) to examine the microstructural nature of their failure surfaces (photomicrographs of these specimens were taken at each stage of this process). In all three types of mechanical specimens, the observed failures were characterized by an absence of fractured fiber glass filaments and bundles. In virtually every case, individual fiber and bundle pullout was the observed mode of failure. The photomicrograph presented in Fig. 8 illustrates the tortuous path which the failure surface follows to avoid filament fracture on the tensile face of the three-point flexure specimens.

To assess the nature of any microstructural variation which might be present in the SMC materials used in this study, a large number of specimens from each material were subjected to resin pyrolysis followed by treatment with a HCl mist spray. As shown in Fig. 9, a layer of individual glass fibers was observed just under the surface of almost all of these specimens. The orientation of these fibers

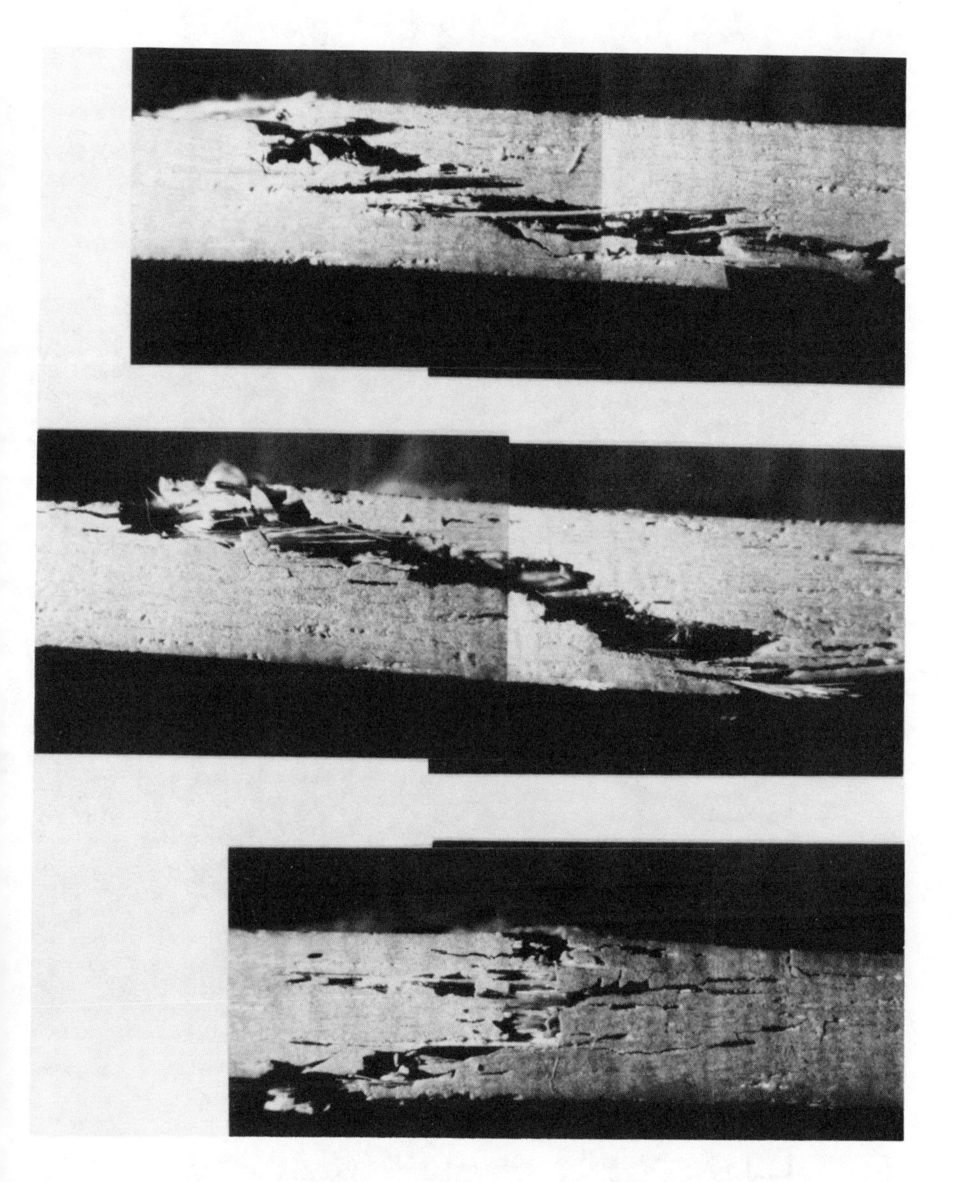

FIG. 7—*Edge view of representative R25 SMC tension strength specimen failures (sample thickness is 3.4 mm, 134 mil), magnification approximately ×5.*

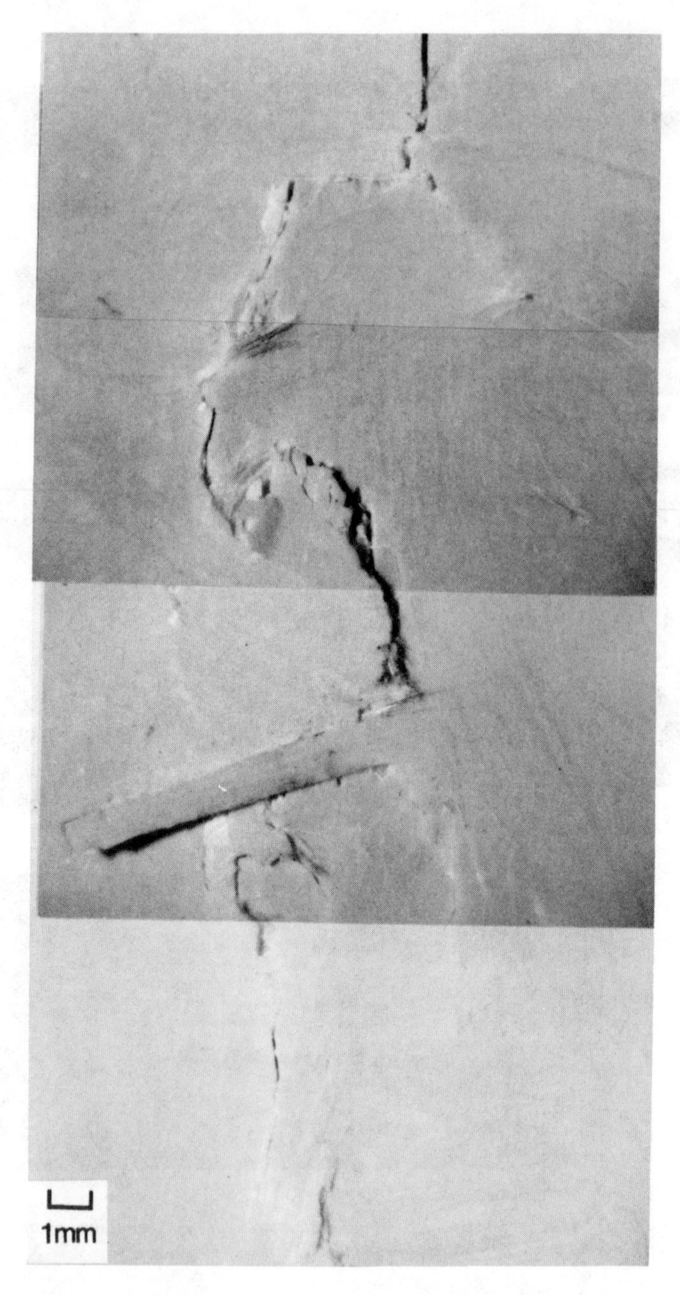

FIG. 8—*A fracture on the tensile face of a three-point flexure specimen.*

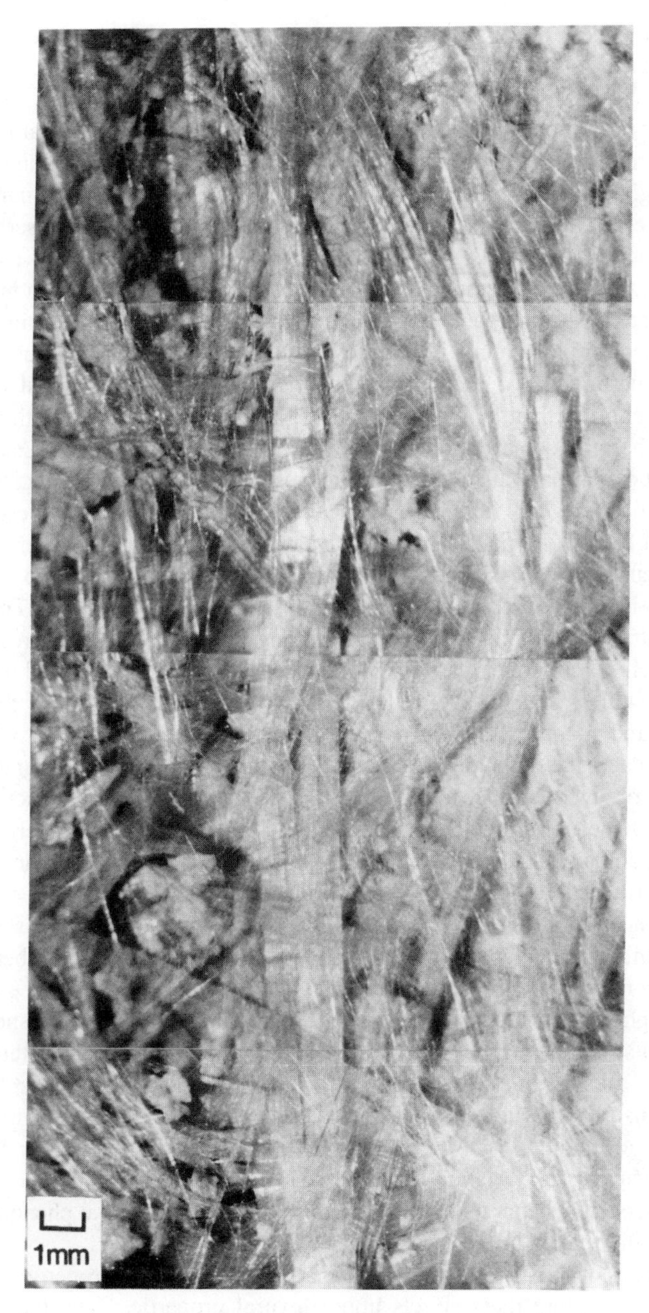

FIG. 9—*Representative surface of a polyester SMC after resin burn-off and treatment with HCl mist spray.*

and the thickness of this surface veil appear to vary erratically across the SMC plaques used in this study. Pyrolysis and HCl mist spray treatment of an entire SMC plaque, 533 by 610 mm (21 by 24 in.), confirmed that these surface veil variations do not correlate with any apparent SMC plaque locational parameter.

In the interior of the two polyester resin SMC materials examined in this study, the fiber glass bundles remained essentially intact (Fig. 10). As shown in Fig. 11, the vinyl ester SMC has a surface fiber glass veil similar to those observed in the polyester SMC materials. However, in the interior of the vinyl ester SMC (Fig. 12), the fiber glass bundles are more effectively dispersed than in either of the polyester SMCs. Other workers [*7,8*][2] have observed somewhat similar SMC microstructural phenomena in both test plaques and structural components.

Photomicrographs of metallurgically polished SMC cross-sectional specimens (Fig. 13) confirm that the vinyl ester SMC is more homogeneous with respect to interior fiber glass distribution than either of the polyester resin SMCs examined in this study. In particular, the polyester resin SMCs have large regions which contain no fiber glass filaments whatsoever. Also, one of the polyester SMCs contained regions of poorly mixed polyester resin and $CaCO_3$ filler (Fig. 14).

As discussed previously in this paper, these microstructural variations can have a significant effect upon the mechanical properties of SMC materials. Three-point flexural strength and modulus would be strongly influenced by the presence of a surface veil of individual glass fibers, its thickness, and any preferred orientation of fibers within this surface veil. In Fig. 15, the data point with the largest strength and modulus came from a three-point flexure specimen which had several intact glass bundles present on its tensile face. These bundles were primarily oriented parallel to the edge of the flexure specimen, which partially resulted in the observed increase in modulus and strength. Furthermore, this specimen contained a significant amount of surface fiber glass veil and the only fractured fiber glass bundle observed in this entire study (Fig. 16).

The concentration of the fiber bundles in the interior of an SMC sheet, their orientation, and the dispersion of the individual filaments from the fiber bundles would strongly influence the four-point flexural properties. As can be observed from the photomicrographs in this paper, the interior of a R25 SMC sheet varies tremendously on both a point-to-point basis and on a material-to-material basis. The vinyl ester SMC, which had the best homogeneity of interior fiber dispersion, also had the lowest coefficients of variation for the four-point flexural strength and modulus data (Table 1). However, even the vinyl ester SMC had large coefficients of variation for its mechanical properties. Thus, the degree of fiber dispersion appears to be only one of many microstructural phenomena that influence the scatter in mechanical properties of SMC.

The effects of SMC microstructural variations upon tensile properties are much more complex than their effects upon flexural properties. The larger testing section of the tension specimens and their catastrophic failure modes make it difficult to analyze the microstructural causes for their failures. Kline [*9*], using

[2]Hagerman, E. M., private communication, 18 May 1978.

FIG. 10—*Representative interior of a polyester SMC after resin burn-off and treatment with HCl mist spray.*

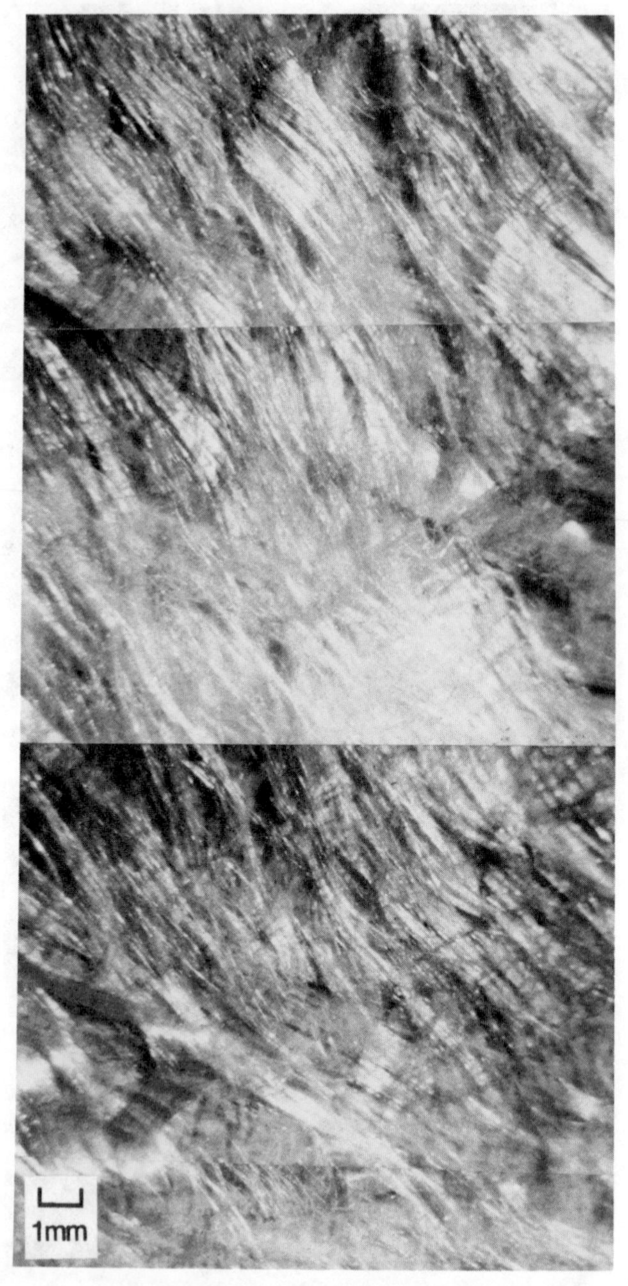

FIG. 11—*Representative surface of the vinyl ester SMC after resin burn-off and treatment with HCl mist spray.*

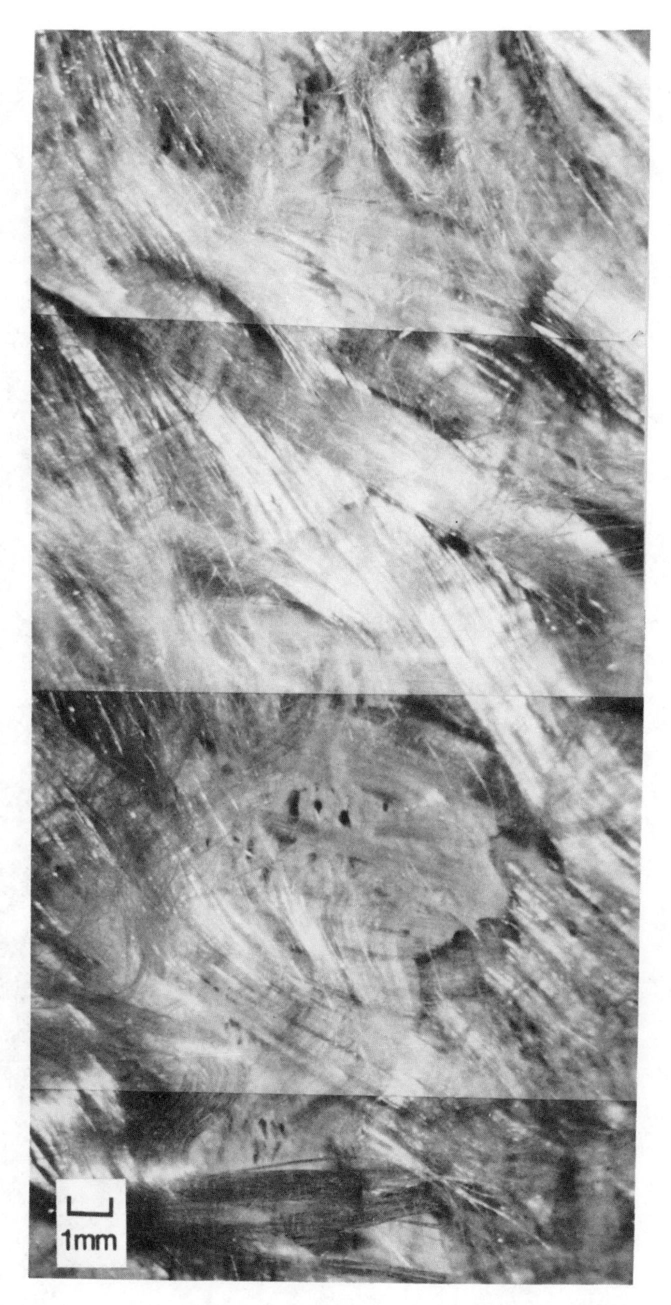

FIG. 12—*Representative interior of the vinyl ester SMC after resin burn-off and treatment with HCl mist spray.*

FIG. 13—*SEM photomontage for a polished cross sectional view of the vinyl ester SMC.*

FIG. 14—*SEM photomontage for a polished cross sectional view of a polyester SMC (the arrows designate some of the regions which contain no $CaCO_3$ filler).*

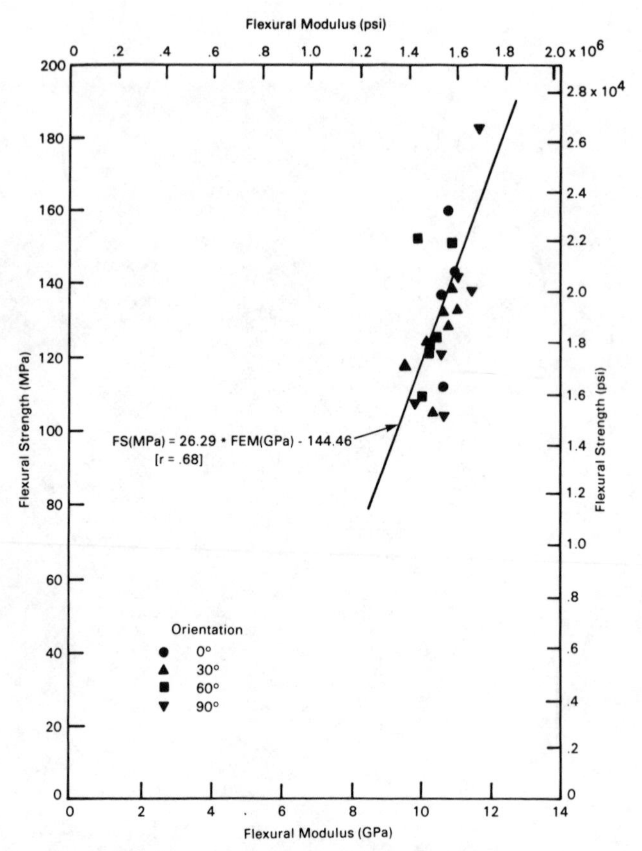

FIG. 15—*Three-point flexural strength versus flexural moduli for a polyester SMC.*

nondestructive testing methods, has recently observed that the failure sites in SMC tension specimens are generally located in small regions of high resin content. While the large resin burn-off specimens used in this study [*4*] indicated an apparent homogeneous distribution of resin, fiber, and filler across all of the SMC plaques, these specimens are much larger than the 9.5 mm (0.37 in.) diameter tensile critical hole size (the upper limit of the tensile flaw site size) observed in these SMC materials.[3] Thus, macrostructural inhomogeneities may play a more important role in determining the ultimate tensile strength than microstructural material variations.

Acknowledgments

The author would like to acknowledge the efforts of the following persons: D. W. Hearn and W. H. Todd of General Motors Advanced Product and Manufacturing Engineering Staff for their help in the fabrication of the test panels; R. P.

[3]Shirrell, C. D., unpublished results.

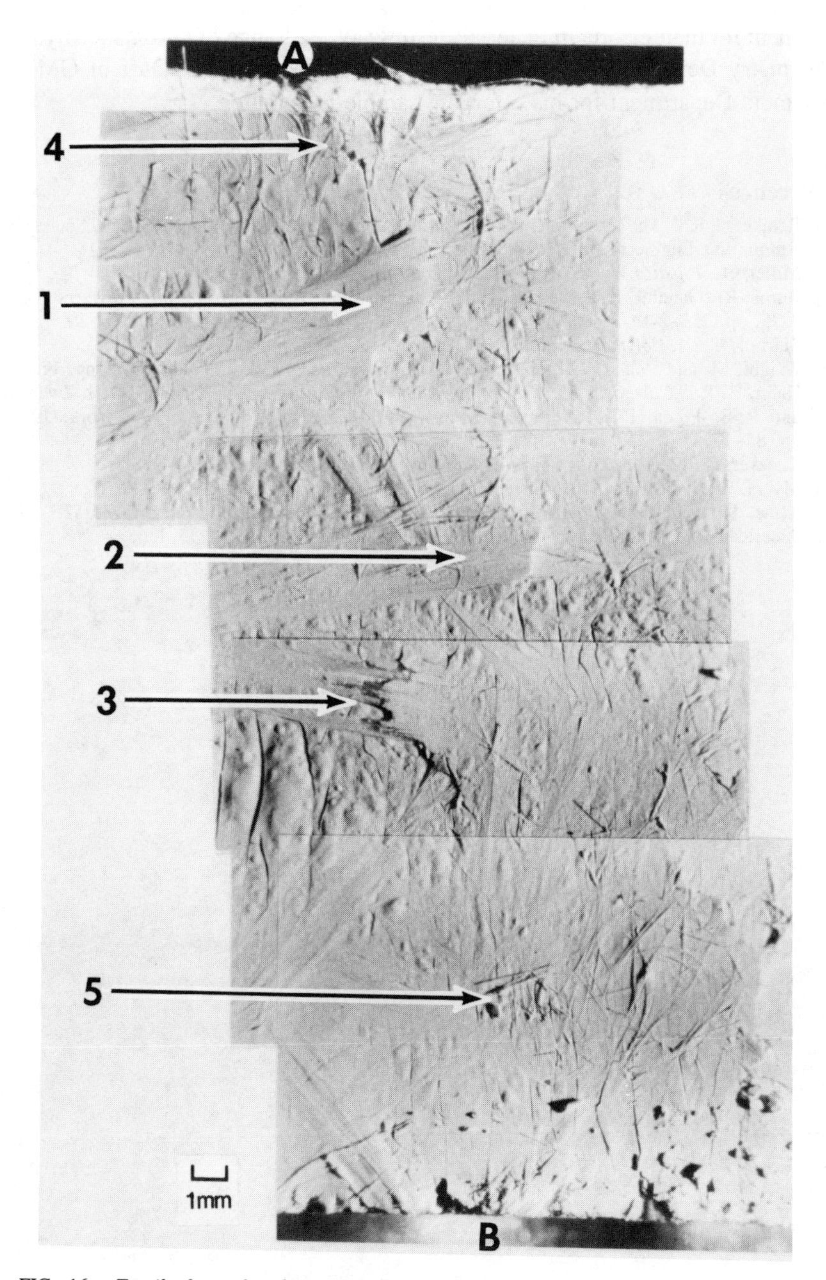

FIG. 16—*Tensile face of a three-point flexural specimen after pyrolysis. Fracture occurred between locations A and B. Fiberglass bundles are indicated at locations 1, 2, and 3. Note fractured fiberglass bundle at location 3 and glass veil at locations 4 and 5.*

Schuler and G. E. Novak of GM Research Laboratories (GMR) Polymers Department for their efforts in materials testing; W. H. Lange of GMR's Analytical Chemistry Department for the SEM photographs; and C. C. Duff of GMR's Polymers Department for his efforts in sample preparation.

References

[1] Reinker, J. K., "The Evolution of SMC in the Automotive Industry," Paper 830356, Society of Automotive Engineers, Inc., Warrendale, PA, 1983.
[2] Miller, B., *Plastics World,* April 1983, pp. 34–39.
[3] Burns, R., Gandhli, K. S., Hanken, A. G., and Lynskel, B. M., *Plastics and Polymers,* Dec. 1975, pp. 235–240.
[4] Shirrell, C. D., *Polymer Composites,* Vol. 4, No. 3, 1983, pp. 172–179.
[5] Knight, M. and Hahn, H. T., *Journal of Composite Materials,* Vol. 9, Jan. 1975, pp. 77–90.
[6] Hahn, H. T. in *Composite Materials in the Automotive Industry,* S. V. Kilkaimi, C. H. Zweben, and R. B. Pipes, Eds., American Society of Mechanical Engineers, New York, 1978, pp. 85–109.
[7] Cheever, G. D., *Journal of Coatings Technology,* Vol. 50, 1978, pp. 36–49.
[8] Myers, F. A. and Han Nyo, *Plastics Compounding,* Sept./Oct. 1981, pp. 83–85.
[9] Kline, R. A. in *Composite Materials: Quality Assurance and Processing, ASTM STP 797,* American Society for Testing and Materials, Philadelphia, 1983, pp. 133–156.

Douglas L. Denton[1] *and Stuart H. Munson-McGee*[2]

Use of X-radiographic Tracers to Measure Fiber Orientation in Short Fiber Composites

REFERENCE: Denton, D. L. and Munson-McGee, S. H., **"Use of X-radiographic Tracers to Measure Fiber Orientation in Short Fiber Composites,"** *High Modulus Fiber Composites in Ground Transportation and High Volume Applications, ASTM STP 873*, D. W. Wilson, Ed., American Society for Testing and Materials, Philadelphia, 1985, pp. 23–35.

ABSTRACT: The degree of fiber orientation is a major factor in determining the mechanical behavior of short fiber composites. A method to determine fiber orientation in short fiber composites such as sheet molding compound (SMC) is presented. A small amount of glass fibers with a high lead content is incorporated into the composite as a tracer. The lead glass tracer fibers mimic the orientation changes that occur during processing. X-radiographs of composites containing the lead glass fibers provide clear, high contrast images of the tracer fibers. Analysis of the X-radiographic image by Fraunhofer light diffraction provides a quantification of the degree of fiber orientation in the composite. Results of fiber orientation and mechanical property measurements for SMC composites molded with different amounts of flow are presented. These techniques are potentially useful for research on fiber orientation, development of prototype parts, optimization of molding conditions, and production quality control.

KEYWORDS: short fiber composites, short fiber orientation, X-radiography, Fraunhofer light diffraction, sheet molding compound, glass fibers, tracer fibers, compression molding, injection molding, reaction injection molding, transfer molding, nondestructive evaluation

The degree of fiber orientation in short fiber composites is a major factor in determining the behavior of the composite. The dependence of composite thermoelastic properties and strength characteristics on short fiber orientation has been emphasized in several recent papers [*1–4*]. The degree of fiber orientation can vary from random to totally aligned. Usually, the degree of fiber orientation

[1]Research and Development Division, Owens-Corning Fiberglas Corp., Technical Center, Granville, OH 43023.

[2]Center for Composite Materials, University of Delaware, Newark, DE 19711. Presently at Research and Development Division, Owens-Corning Fiberglas Corp., Technical Center, Granville, OH 43023.

is between these two extremes. Increasing fiber alignment results in greater anisotropy of the composite.

The orientation state in short fiber composites is determined by processing. Nonrandom fiber orientation is usually an unintentional result of the processing. The primary cause of fiber orientation in compression, injection, reaction injection, or transfer molded composites is flow of the compound during molding. This flow can result in different orientation states at different locations throughout the molded part [5].

Efficient design, prototyping, and production of short fiber composite parts require a knowledge of the fiber orientation state in the part. Knowledge of the specific fiber orientation state in molded parts is also a prerequisite to understanding how this orientation is induced by the processing. Presently, there is no rapid, quantitative method to determine the degree of fiber orientation in short fiber composites.

This paper describes nondestructive techniques for measuring fiber orientation that appear to be feasible for use in research and development, in part prototyping, and in quality control. These techniques are illustrated using sheet molding compound (SMC), but the principles apply to any short fiber composite materials or processing. This technique involves two discrete steps: the creation of an X-radiographic image of a composite containing a tracer fiber and the analysis of the image to mathematically quantify the tracer fiber orientation.

X-Radiographic Tracers

X-radiographs are produced by placing the composite part between a sheet of photographic film and an X-ray source. The closer the film is to the object being examined, the better the resolution of the X-radiographic image. X-radiography offers several advantages over direct visual observation of short fibers in glass reinforced polymer composites. X-radiography is nondestructive and allows glass fibers to be seen through the entire thickness of a molded part, even if the part is optically opaque. Unfortunately, the X-ray absorption of E-glass fibers is not sufficiently greater than the surrounding polymeric matrix to provide high contrast X-radiographs. This is particularly true if the matrix contains an inorganic filler such as calcium carbonate. In addition, composites used in more structurally demanding applications contain higher concentrations of glass fibers. The extensive overlapping of fibers in these composites prevents the clear observation of individual fibers.

The use of X-radiographic tracer fibers is one approach to solving these problems. Tracer fibers which are more X-ray absorbing than E-glass and which are present in low concentrations in the composite can provide high contrast X-radiographs of individual fibers. To be useful, a tracer fiber must accurately mimic the behavior of the E-glass fibers during the compounding and molding of the composite material. Minor differences between tracer and E-glass fibers are probably unimportant if the tracer is used in sufficiently low concentration.

During the initial stages of this work, small diameter (0.25 mm) metal wires were used as tracer fibers. Although the metal fibers provided excellent X-radiographic images, they did not appear to adequately simulate the behavior of the glass fibers during the molding process. This approach was abandoned in favor of the use of glass fibers containing heavy metal ions. A number of glass formulations were considered and several were evaluated. The most successful approach was to use a glass fiber that contained a high concentration of lead.

Lead glass fibers were produced in a pilot plant facility at the Owens-Corning Fiberglas Technical Center in Granville, Ohio. These fibers were produced to be as similar as possible to fibers in a standard E-glass roving (OCF 433) which is used in SMC. The lead glass fibers had the same filament diameter, number of filaments per bundle, and chemical surface treatment as the standard E-glass product. However, two differences did exist between the two fibers. The lead glass had a higher specific gravity than E-glass (4.9 and 2.6, respectively). And the lead glass fiber bundles were not combined to form a roving.

The lead glass was incorporated into SMC using normal processing. One lead glass tracer fiber bundle was fed into the SMC machine chopper with each OCF 433-113 E-glass roving. Since this particular roving nominally contains 64 fiber bundles, the ratio of lead glass to E-glass yielded an SMC in which 1.5 volume % of the glass was tracer fiber. This provides a reasonable concentration of tracer fibers for molded SMC parts which contain 30 to 65 weight % glass fibers and which are about 3 mm thick. The use of lead glass fibers in SMC composites results in X-radiographs with excellent images of the tracer fibers.

Image Analysis

The second step in determining the degree of fiber orientation is analyzing the X-radiographs. Typically, only a single parameter is used to quantitatively describe the degree of fiber orientation [*1*]. One such descriptor is Herman's orientation parameter [*6*]. For planar fiber distributions as in SMC composites, this parameter, f, is given by

$$f = 2\langle \cos^2\Theta \rangle - 1 \tag{1}$$

with

$$\langle \cos^2\Theta \rangle = \int_{-\pi/2}^{\pi/2} n(\Theta) \cos^2\Theta \, d\Theta \tag{2}$$

where

brackets $\langle \, \rangle$ indicate an averaged quantity,
$n(\Theta)$ is the normalized fiber distribution function, and
Θ is the angle between the fibers and principal orientation direction (the direction at which the greatest number of fibers are oriented).

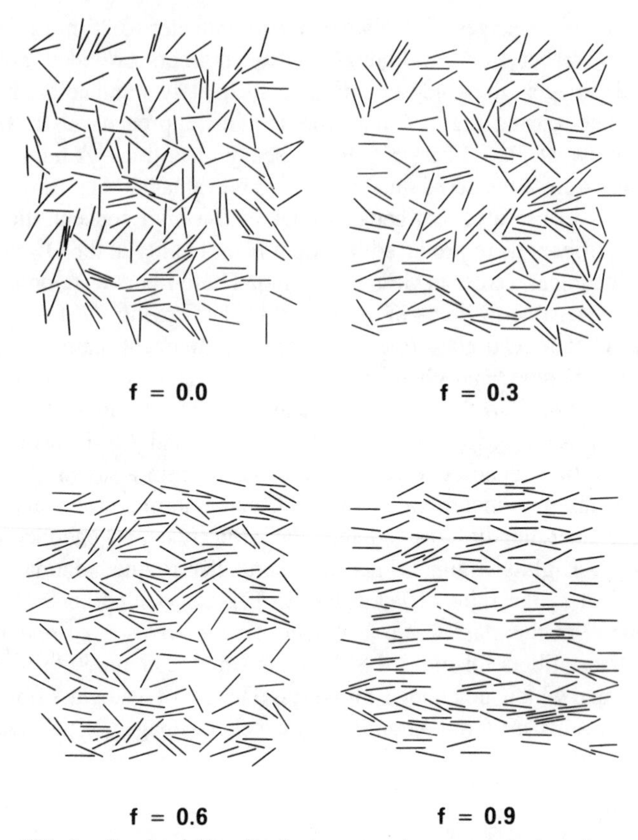

FIG. 1—*Simulated fiber distributions at various states of orientation.*

The values of f range from 0 to 1. A value of f equal to 0 corresponds to a planar random distribution; a value of 1 corresponds to totally aligned fibers. Simulated fiber distributions for various states of orientation are shown in Fig. 1. These distributions were drawn using a Monte Carlo simulation of specified orientation states.

Techniques to determine the degree of fiber orientation can be divided into two categories: those which provide a detailed description of the orientation of each individual fiber and those which provide direct measurement of an orientation parameter. Techniques in the first category include manual measurement of the individual fibers and computer-aided image analysis. To develop a single microstructural parameter to quantify the fiber distribution, these detailed data must be analyzed or averaged in some manner.

The second category of measurement techniques directly provides the microstructural descriptor. Examples of such techniques include Fraunhofer light diffraction, polarized light analysis, and subjective visual assessment of the degree of fiber orientation. These techniques have the advantage of being faster or less

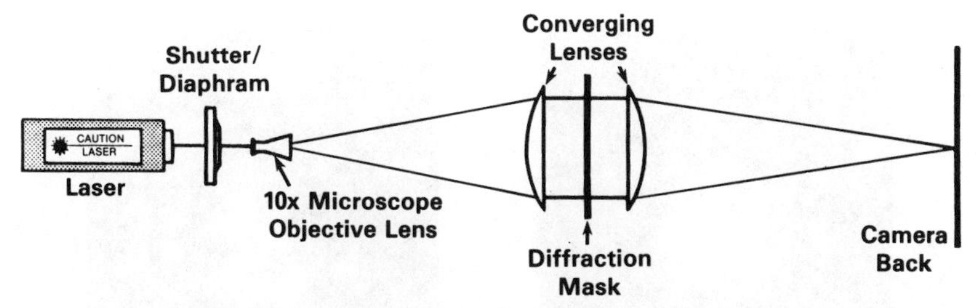

FIG. 2—*Schematic diagram of a Fraunhofer diffractometer.*

expensive than those in the first category, but provide less information about the fiber distribution and may require a calibration method.

Fraunhofer light diffraction was selected for the method of image analysis in this work. When parallel light passes through a diffraction mask, a pattern is produced which is unique to that mask. In this case, the diffraction mask was photographic film on which the image of the fibers was recorded. In the image, the fibers appear transparent on an opaque background. A schematic diagram of a Fraunhofer diffractometer is shown in Fig. 2. The optics in the diffractometer provide parallel light which passes through the diffraction mask and then focuses the diffraction pattern onto a flat surface. This surface can be a recording device such as photographic film.

In this study, a 2.5-mW helium-neon laser was used as a light source. A ×10 microscope objective lens was used as an inexpensive beam expander. Parallel light rays were obtained using a plano convex lens 49 mm in diameter with a focal length of 651 mm. The light was focused on a Polaroid camera back, model 545, using an identical plano convex lens. Polaroid 55 positive/negative film was used to record the diffraction pattern. Exposure time was controlled using a manual shutter located between the laser and the microscope objective lens. The components were mounted on a 1½-m optic bench using standard optical components.

The only instrument variables which affect the photographic image of the diffraction pattern are the size of the shutter aperture and the exposure time. The shutter aperture was set so that it did not block the central portion of the laser beam. The correct exposure time depended on the fraction of the diffraction mask which was transparent. The higher the percentage of the mask which was transparent, the shorter the exposure time. Typically, only one or two shutter speeds provided an image which was neither underexposed (that is, only the central portion of the diffraction pattern appeared) nor overexposed (that is, the detail near the center was lost). For most of the diffraction patterns recorded, exposure times between 0.1 and 1.0 s were typical. Details of the apparatus and its use are given by McGee [7].

Photographic images of the simulated fiber distributions shown in Fig. 1 were used as diffraction masks. The resulting diffraction patterns are shown in Fig. 3. The following features of the diffraction patterns should be noted. As the degree

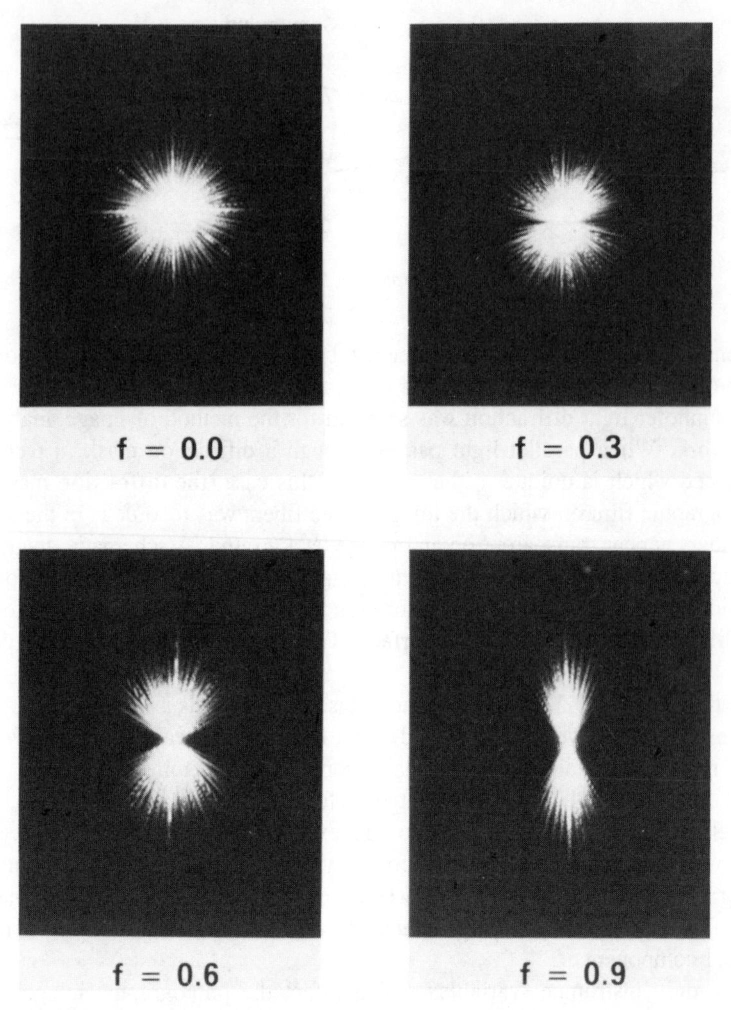

FIG. 3—*Diffraction patterns produced from diffraction masks made from the simulated distributions in Fig. 1.*

of orientation increases, two "lobes" appear in the diffraction pattern. The size of these lobes corresponds to the degree of orientation while the direction of the lobes corresponds to the principal orientation direction rotated by 90°. A complete theoretical analysis of the diffraction pattern is given by McGee and McCullough [*8*].

The diffraction patterns in Fig. 3 can be compared to diffraction patterns from composites with unknown tracer fiber distributions. Through this comparison, the orientation parameter for the composite can be estimated. However, to compare the orientation state between known and unknown specimens, one feature of the diffraction masks must be kept as similar as possible. The major variable besides orientation state that affects the diffraction pattern is the width of the fiber image in the diffraction mask. For the current apparatus, a constant width of 0.03 mm was used.

The need to compare diffraction patterns of specimens with known and unknown fiber orientations is due to inherent limitations of the photographic process. Replacing the photographic film by a device to directly measure the intensity of the diffraction pattern as a function of position would eliminate the need to compare diffraction patterns.

Experimental Evaluation of Methods

SMC-R50 containing lead glass tracer fibers was compounded, matured, and molded into 457 by 305 by 3-mm flat sheets according to a formulation and procedures described by Denton [9]. Panels were molded with three different charge patterns to induce different degrees of fiber orientation. These charge patterns covered 97, 65, and 32% of the mold surface (Fig. 4). The second and third charge patterns produced unidirectional flow of the SMC, which induced substantial fiber orientation. The number of plies in the charge was varied to maintain a constant charge volume.

X-radiographs were made from panels molded with each charge pattern. A Baltograph 5-50 X-ray machine with a tungsten anode tube and beryllium window was used as the radiation source. A 50-kV voltage and 10-mA tube current were used to produce X-radiographs on Kodak Industrex M film exposed for 60 s at a distance of about 840 mm. The intensity of the X-ray source and the exposure time were adjusted to obtain maximum contrast between the tracer fibers and the remainder of the composite. This resulted in different exposure conditions than "normal" X-radiography where the objective is to obtain contrast between the E-glass fibers and the matrix. A wide range of X-ray sources, photographic films, and exposure conditions will probably produce acceptable radiographs with the lead glass tracer fibers.

Differences in the degree of fiber orientation for the panels molded with different charge patterns is evident in prints of the X-radiographs (Fig. 5). The X-radiographs also reveal that fibers subjected to more flow exhibit less bundle integrity. Contact prints made from the X-radiographs were photographically reduced in size with a 35-mm camera. The 35-mm negatives served as the diffraction masks for the Fraunhofer light diffraction experiments.

The diffraction patterns show a distinct elongation as the extent of SMC flow increases (Fig. 6). Orientation parameters were determined by comparisons with standards established from simulated fiber distribution (Figs. 1 and 3). Charge patterns with 97, 65, and 32% mold coverage produced panels with measured fiber orientation parameters of 0, 0.2 to 0.3 and 0.4 to 0.5, respectively.

To illustrate the influence of microstructure on the elastic properties of SMC, the Young's modulus and Poisson's ratio were measured on three panels for each

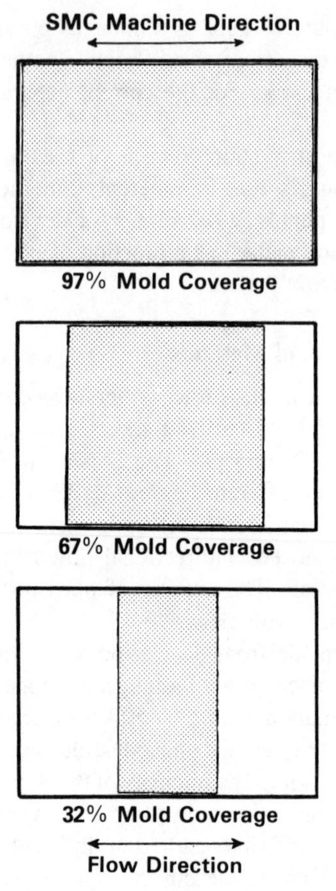

FIG. 4—*SMC charge patterns with 97, 65, and 32% mold coverage.*

of the flow conditions. Tension specimens were cut both parallel and perpendicular to the flow direction (Fig. 7). The two areas of the panel sampled with this layout are nominally equivalent through symmetry. Tests were performed at ambient temperature in accordance with ASTM Test Method for Tensile Properties of Plastics (D 638-82a) using Type I specimens. Strain measurements were made with strain gage extensometers.

Measured values of Young's modulus and Poisson's ratio reflect the degree of fiber orientation in the direction of flow. Mean values for Young's modulus parallel and perpendicular to the flow direction are plotted against measured fiber orientation parameter in Fig. 8. Estimated standard deviations based on twelve observations are indicated by brackets. A similar plot of Poisson's ratio versus measured orientation parameter is shown in Fig. 9. Both Young's modulus and Poisson's ratio exhibit linear changes with the fiber orientation parameter. This type of behavior is expected over the range of fiber orientations observed [7].

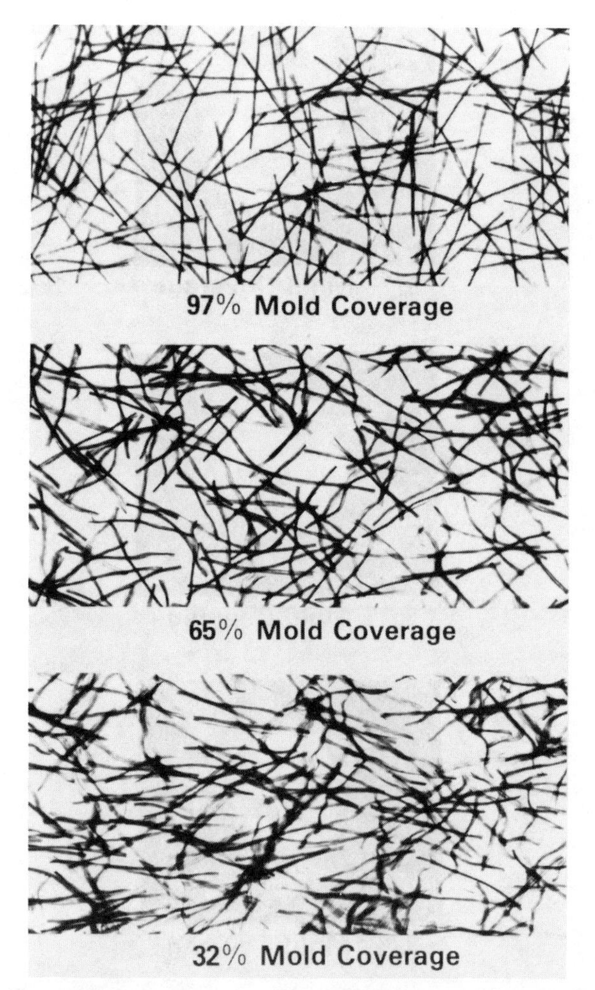

FIG. 5—*Prints of X-radiographs of SMC-R50 containing lead glass tracer fibers and molded with charge patterns in Fig. 4.*

Slight differences in Young's moduli and Poisson's ratios between the two directions in the panels molded with essentially no flow (97% mold coverage) are statistically significant at the 0.90 probability level. These differences are probably due to fiber orientation induced by the SMC machine [*4*]. This slight degree of fiber orientation was not observed in the diffraction analysis. This suggests that improvements in the method of recording the diffraction pattern are needed.

Conclusions

One of the primary parameters for describing the microstructure of short fiber composite materials is the degree of fiber orientation. X-radiography combined

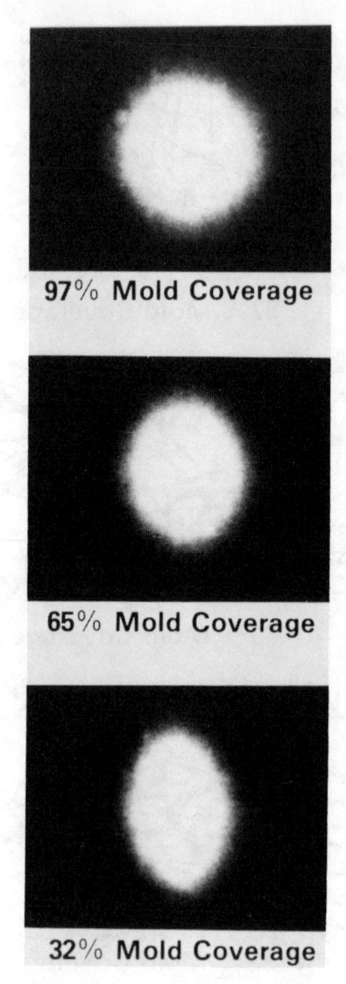

FIG. 6—*Diffraction patterns produced from diffraction masks made from the X-radiographs in Fig. 5.*

with light diffraction techniques can be used to measure fiber orientation. To produce satisfactory X-radiographs, a tracer fiber which is more X-ray absorbent than standard E-glass fibers is required. Special fibers produced from a glass with high lead content serve very well as tracer fibers and produce excellent X-radiographic images.

Fraunhofer light diffraction is a useful technique to analyze X-radiographic images. The shape of the diffraction patterns are a function of the degree of fiber orientation. Random fiber distributions result in circular diffraction patterns. As the degree of fiber orientation increases, two lobes develop in the diffraction pattern. The extent of elongation of the lobes corresponds to the degree of fiber orientation.

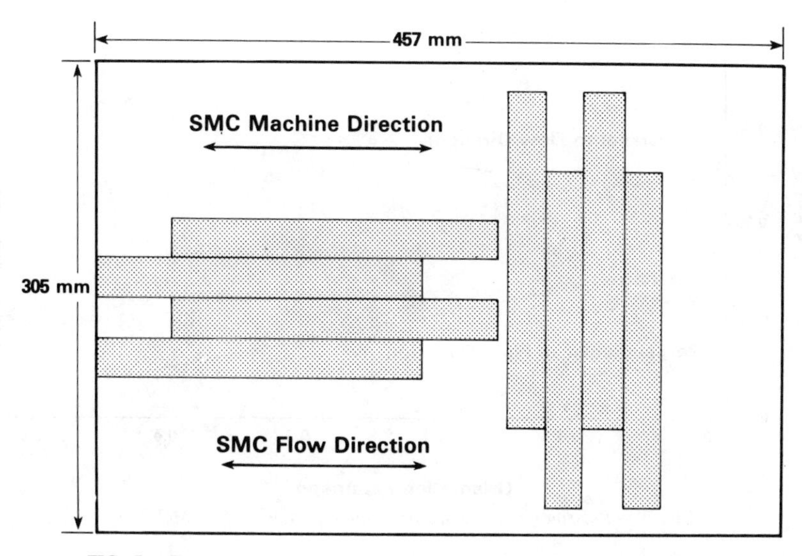

FIG. 7—*Tension specimen sampling pattern for molded SMC-R50 panels.*

Anisotropy in SMC was experimentally correlated to flow-induced fiber orientation. SMC panels were molded with different charge patterns to produce three different fiber orientation states. Fiber orientation parameters measured by X-radiography of lead glass fibers and Fraunhofer light diffraction were directly proportional to measured mechanical properties.

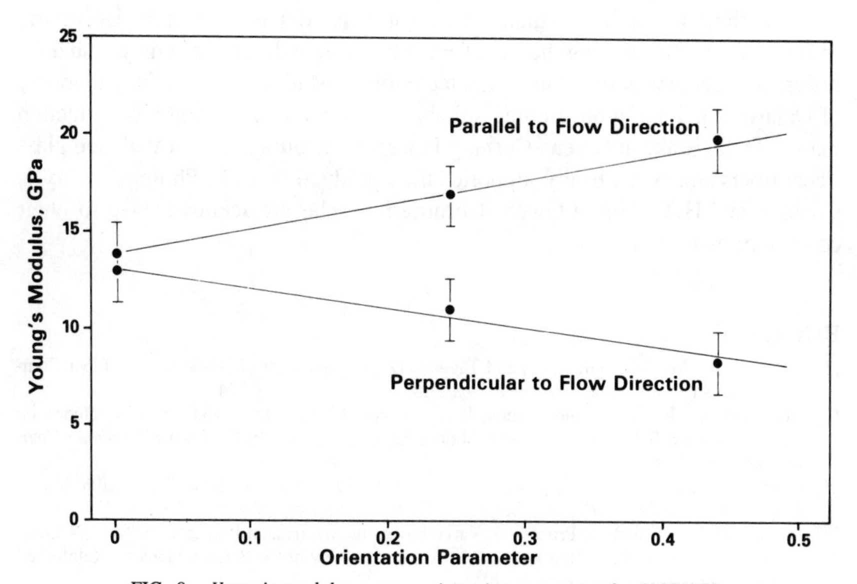

FIG. 8—*Young's modulus versus orientation parameter for SMC-R50.*

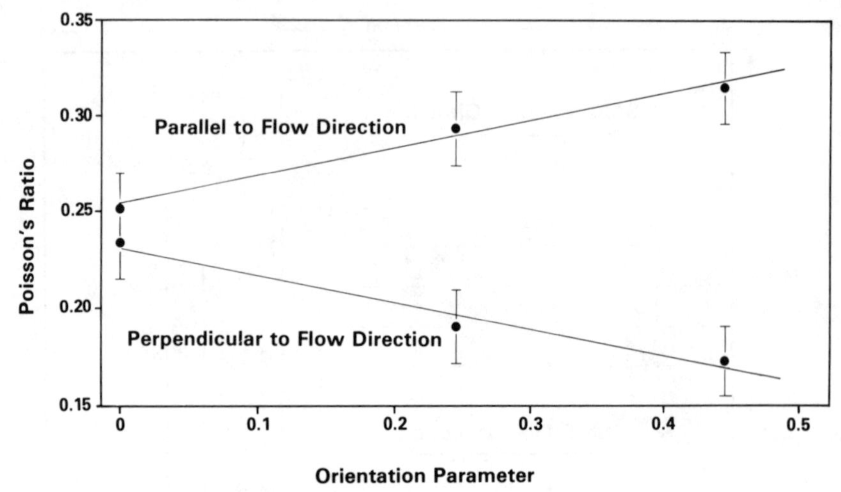

FIG. 9—*Poisson's ratio versus orientation parameter for SMC-R50.*

The results from these preliminary studies indicate that the use of lead glass tracer fibers and Fraunhofer diffraction can be combined to be a powerful tool for measuring fiber orientation in short fiber composites. With additional advances in techniques, this procedure could be adapted to a wide variety of uses to further advance the use of short fiber composite materials.

Acknowledgments

The authors would like to thank Professor R. L. McCullough at the University of Delaware for suggesting the use of Fraunhofer light diffraction analysis and for his encouragement in developing the techniques. Mark Deshon at the University of Delaware helped in developing suitable techniques for making the diffraction masks. D. L. Blake at Owens-Corning Fiberglas encouraged the use of lead glass tracer fibers and W. C. Brady supported their production. L. E. Philipps, Jr., B. J. Creager, and H. C. Gill at Owens-Corning Fiberglas are acknowledged for their contributions to this work.

References

[*1*] Pipes, R. B., McCullough, R. L., and Taggart, D. G., "Behavior of Discontinuous Fiber Composites: Fiber Orientation," *Polymer Composites,* Vol. 3, 1982, p. 34.

[*2*] Masoumy, E., Kacir, L., and Kardos, J. L., "Effect of Fiber-Aspect Ratio and Orientation on the Stress-Strain Behavior of Aligned, Short-Fiber-Reinforced, Ductile Epoxy," *Polymer Composites,* Vol. 4, 1982, p. 64.

[*3*] Chen, C.-Y. and Tucker, C. L., "Mechanical Property Predictions for Short Fiber/Brittle Matrix Composites," *Technical Papers,* Society of Plastics Engineers, Vol. 29, 1983, p. 337.

[*4*] Denton, D. L., "Effects of Processing Variables on the Mechanical Properties of SMC-R Composites," in *Proceedings,* 36th Annual Conference, Society of the Plastics Industry, Reinforced Plastics/Composites Institute,16-A, 1981.

[5] McCullough, R. L., Pipes, R. B., Taggart, D., and Mosko, J. in *Composite Materials in the Automotive Industry,* American Society of Mechanical Engineers, New York, 1978, p. 141.
[6] McCullough, R. L. in *Treatise on Material Science and Technology,* Vol. 10, Pt. B, Academic Press, New York, 1977, p. 453.
[7] McGee, S. H., "The Influence of Microstructure on the Elastic Properties of Composite Materials," Ph.D. dissertation, Dept. of Chemical Engineering, Univ. of Delaware, Newark, DE, 1982.
[8] McGee, S. H. and McCullough, R. L., *Journal of Applied Physics,* Vol. 55, 1984, p. 1394.
[9] Denton, D. L., "Mechanical Properties Characterization of an SMC-R50 Composite," in *Proceedings,* 34th Annual Conference, Society of the Plastics Industry, Reinforced Plastics/Composites Institute, 11-F, 1979.

John F. Mandell,[1] *Fredrick J. McGarry,*[1] *and Chia-Geng Li*[1]

Fatigue Crack Growth and Lifetime Trends in Injection Molded Reinforced Thermoplastics

REFERENCE: Mandell, J. F., McGarry, F. J., and Li, C.-G., **"Fatigue Crack Growth and Lifetime Trends in Injection Molded Reinforced Thermoplastics,"** *High Modulus Fiber Composites in Ground Transportation and High Volume Applications, ASTM STP 873,* D. W. Wilson, Ed., American Society for Testing and Materials, Philadelphia, 1985, pp. 36–50.

ABSTRACT: Glass and graphite fiber reinforced injection molded thermoplastics contain very short fibers, resulting in matrix/interface dominated properties. Molded parts typically fail by propagation of a single dominant crack. A fiber avoidance crack growth model is presented in which the crack extends by the coalescence of isolated failure zones; the crack tip process zone dimension is fixed by the microstructure (fiber) size and so does not change with test conditions or K_{max} in fatigue crack growth. Fracture toughness data for several materials correlate with this model. Fatigue crack growth follows a power law relationship with an exponent close to 4 for the pure matrices and 7 to 8 for the composites. The threshold K_I values where cracks will not grow are a much higher fraction of the fracture toughness for the composites than for most unreinforced polymers or metals. The trends of *S-N* curves for unnotched molded bars follow the same exponent as observed in fatigue crack growth. Shifts in the pure matrix failure mode in fatigue are reflected in the local failure mode of the composites.

KEY WORDS: short fiber composite materials, fracture, fatigue, crack propagation modes, injection molding

A number of glass and graphite fiber reinforced thermoplastic compounds intended primarily for use in injection molding processes are currently available. While the thermoplastic matrices span the range from inexpensive to high performance, much of the engineering interest centers on recently developed strong, temperature-resistant polymers such as polyamide imide (PAI), polyether imide (PEI), and polyetherether ketone (PEEK), as well as engineering plastics which have been available for a longer time, including nylon 6,6 (N66), polycarbonate (PC), polysulfone (PSUL), and polyphenylene sulfide (PPS). Regardless of the

[1]Principal research associate, professor, and graduate student, respectively, Department of Materials Science and Engineering, Massachusetts Institute of Technology, Cambridge, MA 02139.

initial fiber length, the injection molding operation (as well as associated extrusion in some cases) degrades the fibers so that their final length seldom reaches 1 mm; typical fiber lengths are closer to 0.2 mm, representing a length to diameter aspect ratio of 20. The fibers are short enough so that failure generally is matrix-dominated for all orientations. Fiber orientation is determined by the melt flow pattern in the mold, varying from point to point and through the thickness. While orientation is never perfect, high levels are often achieved in local regions. The fibers are well dispersed and the material is more uniform than typical sheet molding compound (SMC) composites made from strands.

The failure pattern of this class of composites is more similar to that of unreinforced materials than is the case with most composites. The preponderance of matrix cracking characteristic of SMC and continuous fiber composites in fatigue is generally absent. Instead, failure is almost always by the development and propagation of a dominant crack. This mode of failure is best studied in a fracture mechanics context; previous papers from this study [*1–4*] have reported on the fracture and fatigue behavior of a variety of commercial materials. This paper presents new fatigue data and also reconsideration of some previously published results.

Experimental Methods

The materials used are commercial injection molding compounds containing 30 to 40% by weight glass or graphite fibers. The graphite fibers are polyacrylonitrile-based with a modulus in the range of 200 GPa (30×10^6 psi); they are referred to as carbon in much of the industry literature, and the terms carbon and graphite will be used interchangeably here. The ultimate tensile strength (UTS) and *S-N* fatigue data were generated on 0.32-cm-thick end-gated [ASTM Test Method for Tensile Properties of Plastics (D 638-82a)] Type I tensile bars, while fatigue crack growth and fracture toughness tests used single edge notched tension specimens cut from 0.64-cm-thick center face gated 10 by 20-cm plaques.

Details of material sources, specimen preparation, testing, and data reduction are contained in earlier papers [*1–4*]. Fiber length distributions for most of the materials are given in Ref *2*. While the fiber orientation varies through the thickness, all data given here are for specimens loaded in uniaxial tension parallel to the dominant fiber orientation; all cracks grow normal to this direction. The effects of fiber orientation have been considered in Ref *1*. A broader discussion of the effects of fiber orientation on general mechanical properties may be found in Refs *5* and *6*. Specimens were conditioned and tests run in an air-conditioned laboratory at approximately 23°C, without humidity control. Of the materials tested only N66 is significantly affected by moisture content at 23°C; measured moisture contents for the N66 specimens were 1.4 to 1.6% by weight of the matrix.

Fatigue tests were sine wave, load controlled, at a minimum to maximum load ratio of 0.10. Frequencies given were chosen to avoid significant hysteretic heating; for *S-N* data sets the frequency was varied inversely with the maximum

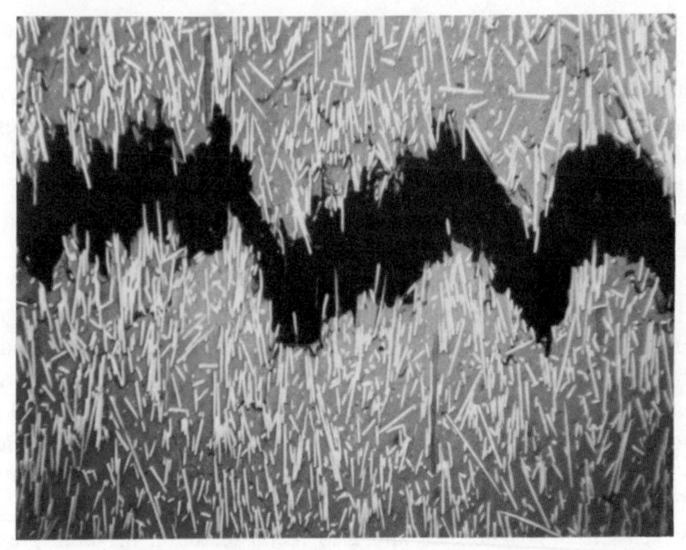

FIG. 1—*Profile of crack in graphite/PC (fiber dia 10 μm), from Ref* 4.

load to maintain an approximately constant average load rate. Single cycle data were obtained in a ramp test at a rate consistent with the fatigue tests.

Results and Discussion

Mode of Crack Growth

The mode of crack propagation has been described previously [*1*,*4*] as a fiber avoidance mode. Figure 1 shows a profile of a crack path on a previously polished surface, and Fig. 2 shows a scanning electron micrograph of a fracture surface. The magnification is evident from the fibers, which are 10 to 11 μm in diameter. These figures are typical of cracks in the entire range of materials discussed here, both in monotonic and fatigue tests. The macroscopic crack grows in a zigzag pattern, passing around most of the fibers. On the fracture surface (Fig. 2) the pattern is similar along the crack and across the thickness. Thus, the crack tip at any instant forms a zigzag pattern across its front, through the thickness.

Details of the process zone at the crack tip have been described for a variety of material systems in Ref *4*. Figure 3 shows the crack tip region for glass reinforced PC. Ahead of the main crack tip, on the scale of the zigzag pattern, is a group of isolated crazes, predominantly originating at fiber ends. Other matrices tend to show a similar appearance, but with isolated shear yielded zones [*4*]. The micrographs were taken on prepolished surfaces and tend to emphasize shear yielding in many cases while crazing may be dominant on the interior, where plane-strain conditions are expected to prevail. The macroscopic crack has

FIG. 2—*Fracture surface at notch in graphite/N66* [4].

FIG. 3—*Crack tip process zone in glass/PC* [4].

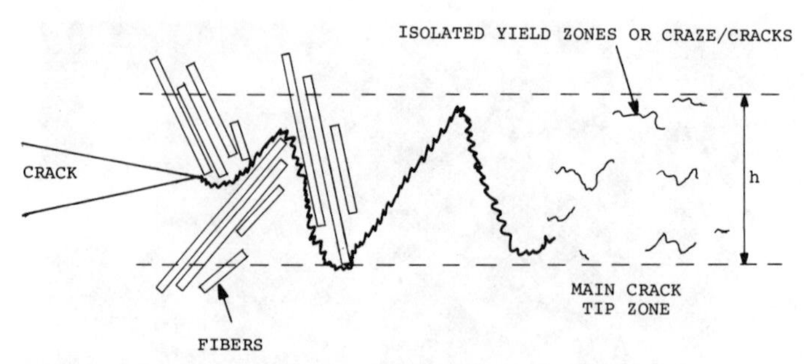

FIG. 4—*Schematic of crack zone in fiber avoidance mode.*

been shown to propagate by the coalescence of the isolated crazes, cracks, and shear yield zones in virtually all materials studied.

Figure 4 is a schematic of this mode of crack extension. It is evident that the fiber avoidance and crack tip coalescence characteristics define a zone size, h, where cracking occurs. This zone size is empirically observed [*1*,*4*] to be defined by the length of local fiber agglomerations as shown in Fig. 4. The microstructure is fixed by the material system and melt flow in the molding process. Thus, h is a parameter associated with this mode of crack extension under all conditions of rate, temperature, fatigue, etc. While more data are needed, microscopy studies to date support the independence of h from the loading conditions and history.

The extension of the main crack tip is not a steady process of zigging and zagging around the fiber agglomerations. Rather, in all material systems studied [*4*], the main crack at a given location through the thickness is stationary at a position where its path is blocked by fibers, or where it is extending in a direction away from the path of the remainder of the crack front. The isolated failure zones ahead of the main crack then extend, coalesce, and form a new main crack path. Thus, main crack extension is associated with the formation and extension of the isolated failure regions ahead of its tip, over the domain roughly defined by h. In some cases isolated failure zones also may be observed beyond the region defined by h; however, for the crack to grow in this mode, it appears only necessary that separation be possible at the imposed K_I value out to the distance h.

Other modes of crack propagation can also be conceived. A common model for very short fiber composites [*7*] involves a planar crack spreading through the matrix, with debonded fibers being extracted from either surface. Micrographs of lower fiber volume fraction composites [*5*,*6*] appear more consistent with this mode than are the materials used in the present study. Only the brittle matrix PPS with poorly bonded fibers showed some evidence of this mode [*4*]. In fact, only incidental fibers are extracted from the matrix in most cases [*4*] (Fig. 1). For the most part the fibers are not well enough aligned to be extracted by an opening of the crack without widespread matrix crumbling or deformation [*1*,*4*].

TABLE 1—*Fracture toughness,* K_Q, *and ultimate tensile strength, UTS.*[a]

Property	Reinforcement	Matrix				
		N66	PC	PSUL	PPS	PAI
K_Q, MN $m^{-3/2}$	None	3.9	2.8	2.9	0.8	...
	Glass	9.9	8.7	6.5	7.0	9.4
	Carbon	9.4	7.5	7.2	6.6	...
UTS, MPa	None[b]	74	72	77	35	140
	Glass	181	161	150	181	203
	Carbon	256	203	197	156	231

[a]After Refs *1* and *2*; K_Q values for unreinforced matrices have been updated since earlier publication and are now valid by ASTM Method for Plane-Strain Fracture Toughness of Metallic Materials (E 399-83). K_Q values differ from those in Figs. 7 and 8 due to differences in rate and test geometry.
[b]Values for N66, PC, and PSUL are yield stress.

Fracture Toughness

Fracture toughness (K_Q) and UTS values have been published previously for a number of material systems, as listed in Table 1 [*1*]. The materials are designated by the fiber and matrix, such as C/N66 for carbon (graphite) reinforced N66; G is used to designate glass fibers. The fracture toughness data are also presented in Fig. 5. The parameter r in Fig. 5 is the calculated radius from the crack tip using linear elastic fracture mechanics approximation; the local stress at the crack tip is approximately [*8*]

$$\sigma_{ij} = \frac{K_I}{\sqrt{2\pi r}} f(\Theta) \tag{1}$$

for isotropic materials where

r and Θ = polar coordinates,
σ_{ij} = local stress components, and
K_I = opening mode stress intensity factor.

Fracture in the fiber avoidance mode is assumed [*1*] to occur when the local stress in the load direction reaches the UTS of the material over the entire zone, h, giving approximately

$$r = h/2 = \frac{1}{2\pi}\left(\frac{K_Q}{\text{UTS}}\right)^2 \tag{2}$$

In Fig. 5 the right-hand side of Eq 2 is compared with the measured microstructure size, taken as the fiber length, l_f*, which is exceeded by 5% of the fibers, a length intended to best represent the size of the small agglomerations of fibers depicted in Fig. 4 and evident in Fig. 1. The agreement shown in Fig. 5 is surprisingly good given the rough approximations and assumptions involved.

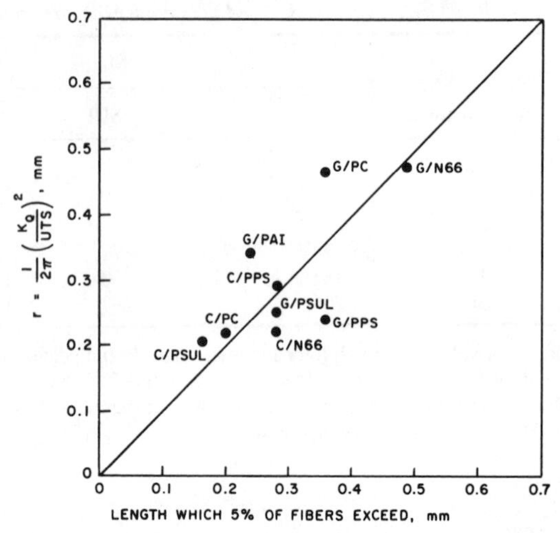

FIG. 5—*Predicted process zone dimension,* r, *versus experimental fiber length characteristic for nine materials* [1].

This yields a prediction of the fracture toughness as

$$K_Q = (2\pi l_f{}^*)^{1/2}\,\mathrm{UTS} \tag{3}$$

This prediction for K_Q is in much better agreement with the data than one based on the fracture toughness of the matrix. As shown in Table 1, the matrix toughness correlates very poorly with composite toughness when the brittle PPS matrix is considered.

Equation 3 predicts that the ratio K_Q/UTS should be constant regardless of rate or temperature conditions, since l_f^* is fixed. A few recent tests on graphite reinforced PC at 25 and 75°C and displacement rates of 0.015, 0.15, 1.5, and 15 cm/s shown in Fig. 6 tend to support this observation. It should be noted that fiber pullout models could also show similar trends; the use of Eq 3 is based on our observations of modes of crack extension [*1,4*] which are limited to this group of materials.

Fatigue Crack Growth

Fatigue crack growth data have been obtained for PSUL and N66 matrix systems over the complete range of crack growth rates from threshold values where no growth can be observed up to maximum K_I (K_{max}) values approaching K_Q. A narrower range of these results for PSUL systems was given in Ref *3*. Figures 7 and 8 give results for PSUL and N66 systems, respectively. The solid lines represent Paris-type power law relationships of the form

$$da/dN = AK_{max}{}^m \tag{4}$$

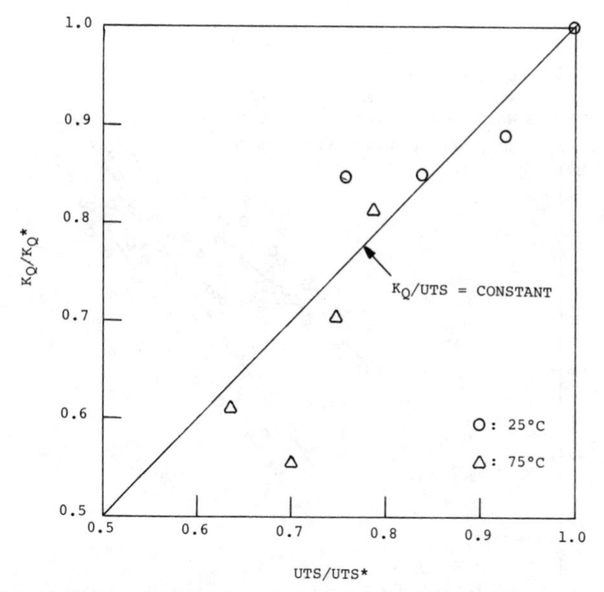

FIG. 6—*Relative changes in* K_Q *and UTS of graphite/PC at 25 and 75°C for displacement rates from 0.015 to 15 cm/s (*K_Q^* *and UTS* are values at 25°C and 15 cm/s).*

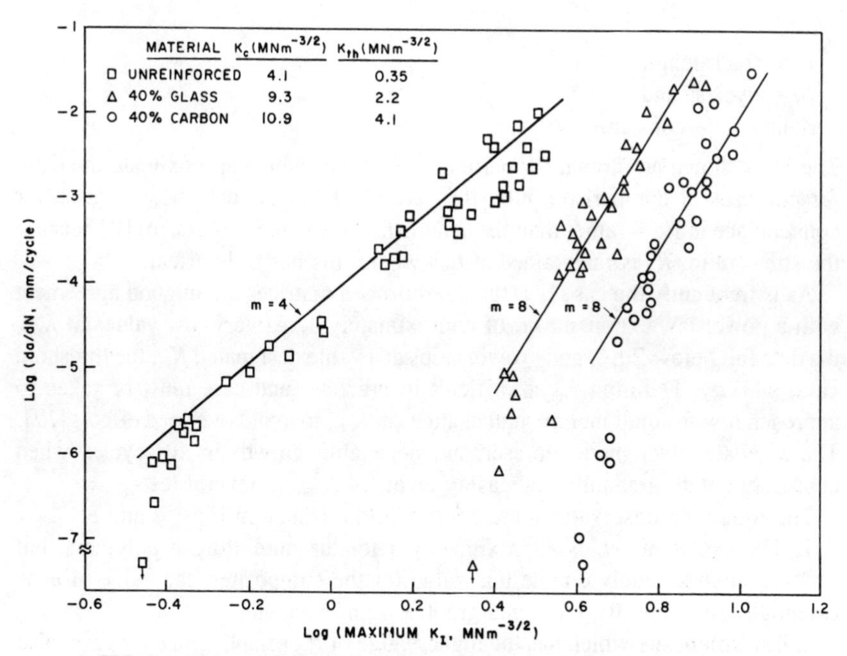

FIG. 7—*Fatigue crack growth data for PSUL matrix systems,* R = *0.1, 10 Hz.*

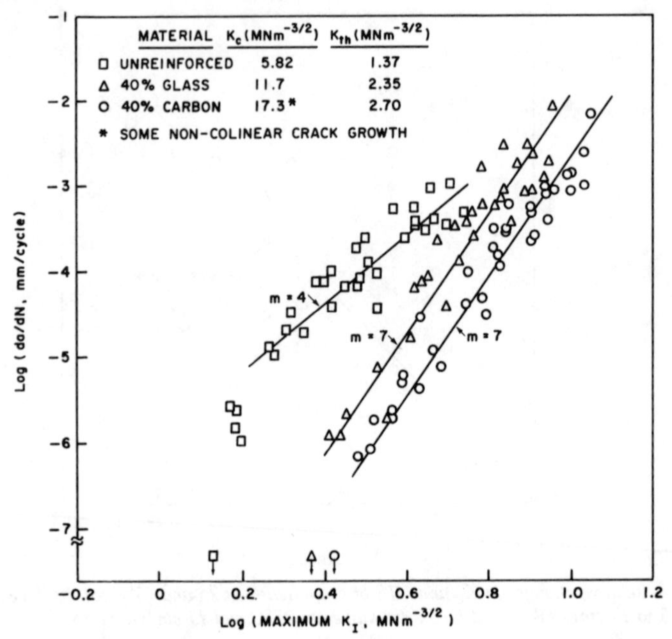

FIG. 8—*Fatigue crack growth data for N66 matrix systems,* R = *0.1, 10 Hz.*

where

a = crack length,
N = cycles, and
A and m = constants.

The lines shown are drawn for particular m values which approximate the data, for purposes of comparison only; they are not fit to the data. K_{max} is used for convenience in Eq 4 rather than the usual ΔK_I (stress intensity range) [*9*] because the stress ratio, R, is maintained at 0.1 where this has little effect.

As is frequently the case [*9*], the unreinforced matrices are in good agreement with a power law exponent, m, of approximately 4. At very low values of K_{max} the data fall below Eq 4, and growth stops at a value designated K_{th}, the threshold value of K_{max}. Defining K_{th} is difficult in practice, and care must be taken to approach it with small incremental changes in K_{max} to avoid overload effects [*10*]. The K_{th} data given here represent no detectable growth in 10^6 cycles when approached with gradually decreasing levels of K_{max} in several tests.

The following observations are evident in the results in Figs. 7 and 8:

1. The exponent, m, is approximately 4 for the unreinforced polymers, but shifts to approximately double this value for the composites; the value of m is essentially the same for glass and graphite reinforcement.

2. The composite which has the higher value of K_Q (graphite in each case) also has the most resistance to fatigue crack growth. The glass and graphite fatigue crack growth curves are separated by approximately a constant amount on the log

scale over the entire range. Thus, the ratio of K_{max}-values for the two reinforcements at a given crack growth rate is approximately constant. This is similar to results for unreinforced polymers with different K_Q's [9] and is true of comparisons of the PSUL and N66 curves as well as those of their composites.

3. The values of K_{th} for the composites are relatively high compared to those of many unreinforced matrices, particularly amorphous polymers like PSUL.

The results for K_{th} are the only known data of this type for injection molded polymers or composites. They show an interesting trend of very high K_{th}/K_Q ratios for the composites. Metals, including steel, aluminum, and titanium alloys, have reported ratios in the range of 0.03 to 0.08 at $R = 0.0$ to 0.2 [10]. The ratio for unreinforced PSUL is 0.085, while for nylon, with a coarse, spherulitic, semicrystalline microstructure, the ratio is 0.23. By contrast, the K_{th}/K_Q ratios for PSUL composites are 0.24 for glass and 0.38 for graphite; for N66 composites the corresponding values are 0.20 and 0.16, similar to the nylon matrix. The absolute values of K_{th} for the composites are in the same range as many aluminum alloys having much higher K_{Ic}-values, in the range of 40 MN $m^{-3/2}$.

The shift in m from around 4 for the matrices to around 7 or 8 for the composites is also significant. Since the fatigue crack growth curves are essentially fixed at the top end by K_Q as with unreinforced polymers [9], a high exponent corresponds to an increase in fatigue resistance; an infinite value of m would represent a complete absence of cyclic effects, with failure occurring at K_Q. In unreinforced materials, high values of m are associated with very brittle behavior and correspondingly low levels of K_Q; examples are highly cross-linked epoxies [9], ceramics, and inorganic glasses [11]. Continuous fiber composites tend to have values of m in the 7 and higher range for interlaminar cracking [12,13] (low K_Q) and 10 to 15 for in-plane, transfiber cracking [14,15] (high K_Q).

It is evident in Figs. 1–4 that cracks in the short fiber reinforced thermoplastics considered here extend primarily in the matrix. The value of m for both matrices in Figs. 7 and 8 is approximately 4, and yet the matrix-dominated composites have a much higher value in each case. The limited data in the literature for similar materials indicate exponents in the range of 7 or 8 for crack growth perpendicular as well as parallel to the dominant fiber direction [16]. The exponent of approximately 4 is in good agreement with data for many unreinforced polymers [9] and some metals [17]. However, when a broader range of test variables is considered than the simple constant amplitude sine wave with $R = 0.1$ used here, the behavior is more complex and m is difficult to predict from basic considerations [18]. As indicated schematically in Fig. 9, the length and thickness of a craze zone (or other type of yield zone) decreases with K_{max}^2 as K_{max} is reduced from K_Q in fatigue [9]. The crack tip opening displacement also decreases with K_{max}^2. These provide an obvious basis for an exponent, m, of 2. Prediction of a higher exponent such as 4 usually requires some consideration of the properties of the yielded material itself [18].

The most obvious basis for an approximate doubling of m in the composites rests with the observation that the zone size, h, in Fig. 4, is fixed by the micro-

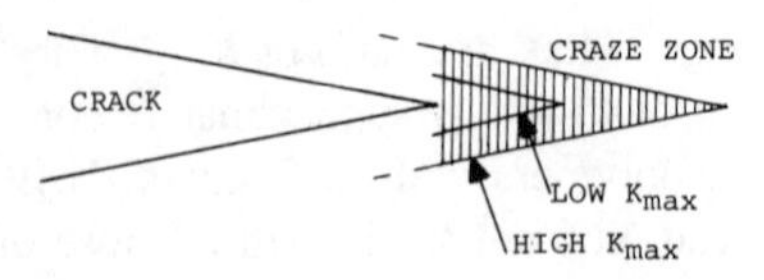

FIG. 9—*Schematic of craze zone of unreinforced polymer during fatigue crack growth at high and low* K_{max} *levels.*

structure and is not observed to change significantly in fatigue crack growth [*4*]. Thus, the isolated craze or shear yield zones ahead of the crack tip must nucleate, grow, and coalesce at the same distance from the main crack regardless of K_{max}, in contrast to the case in Fig. 9. While a theoretical solution for the geometry in Fig. 4 is not available, as K_I decreases the isolated zones must experience a more rapidly decreasing stress and strain field than does the zone surrounding the crack tip stress singularity in Fig. 9. Another factor present with the composites is that the isolated crazes in Figs. 3 and 4 are usually associated with fiber ends. As they grow, there must be a reduction in the load carried by the fiber, which would affect the individual craze growth and also render the material in zone h in Fig. 4 much softer, since most of the load transfer into the fibers will presumably have been eliminated.

The crack growth model in Fig. 4 also provides some rationalization for the high K_{th}-values. Since main crack propagation requires that the isolated craze/yield zones nucleate and grow at a fixed distance from the main crack tip, they do not experience the singularity which always provides a very high stress as the distance collapses on the crack tip, even at very low K_{max}-values. If isolated failure zones cannot nucleate at a distance $h/2$ from the main crack tip at a particular value of K_{max}, then the main crack cannot grow, at least in this coalescence mode.

S-N *Fatigue*

Generally, it is not possible to predict the lifetime of an uncracked specimen of metal from fatigue crack growth data. The crack initiation part of the lifetime, usually sensitive to such factors as surface finish, complicates the prediction. The lifetime may be predicted for a specimen containing an initial notch or flaw which begins growing on the first cycle and grows to a length where it becomes critical (K_c). Equation 4 can be integrated from the initial to the critical crack length (at the imposed cyclic stress) to give an *S-N* curve trend [*17*]

$$N \alpha S_{max}^{-m} \tag{5}$$

where

S_{max} = maximum cyclic uniaxial tensile stress, and
N = the cycles to failure.

(For values of R not close to zero, ΔS should be used in Eq 5.) Thus, if the specimen lifetime is dominated by crack growth following Eq 4, then the

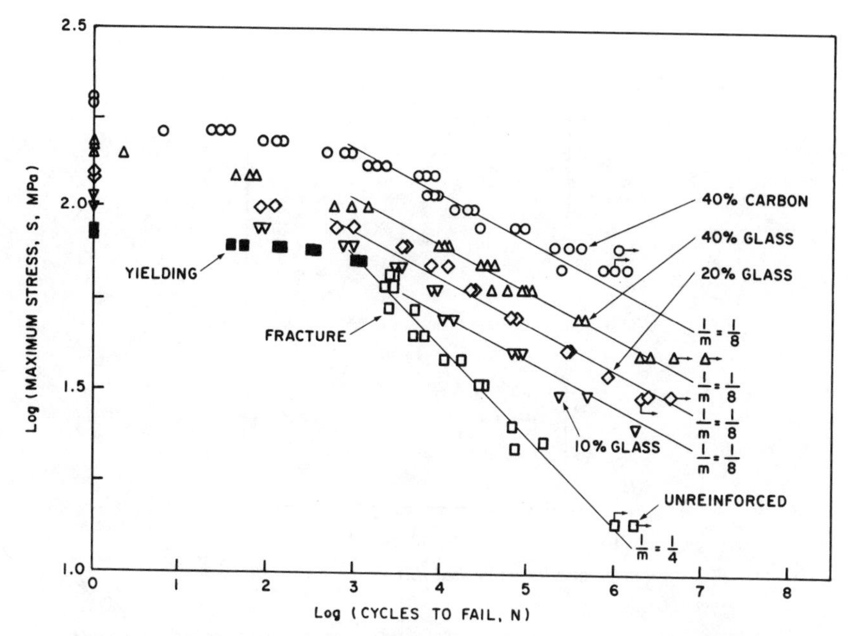

FIG. 10—S-N *data for unnotched PSUL matrix systems,* R = *0.10, 5 to 20 Hz* [3].

specimen *S-N* curve should show the same exponent, *m*, on a log-log plot of *S* versus *N*.

Figure 10 is such a plot for a range of PSUL matrix composites, from Ref *3*. In the portions of the data with a straight line shown for comparison (not curve fit), evidence of growing cracks during fatigue was indeed found, while the low cycle parts of the curves failed suddenly, apparently by matrix yielding [*3*]. The pure PSUL showed analogous behavior, with yielding (necking) at low cycles and crack growth at high cycles. Thus, the matrix failure mechanism was observed in Ref *3* to be present in the composites as well. Comparison of Figs. 7 and 10 also shows that unnotched *S-N* curves do have trends with the same exponent as in fatigue crack growth, for both the matrix and the composites (the slope is $-1/m$ because the coordinates are inverted between Figs. 7 and 10). Less complete data are available on other systems, but the same trends appear to hold when failure is by crack growth; Fig. 11 gives analogous *S-N* results for systems with a PEI matrix which is amorphous, like PSUL. PC systems show similar trends. The unnotched nylon matrix materials fail by matrix yielding, not stable crack growth, over the entire lifetime range [*3*]. However, when the nylon materials are notched, they behave much like the amorphous matrix systems.

Discussion

The results in the last section suggest that the lifetime trends of molded parts can be predicted directly from fatigue crack growth data in some cases. The strong implication is that the materials contain flaws which are growing over most

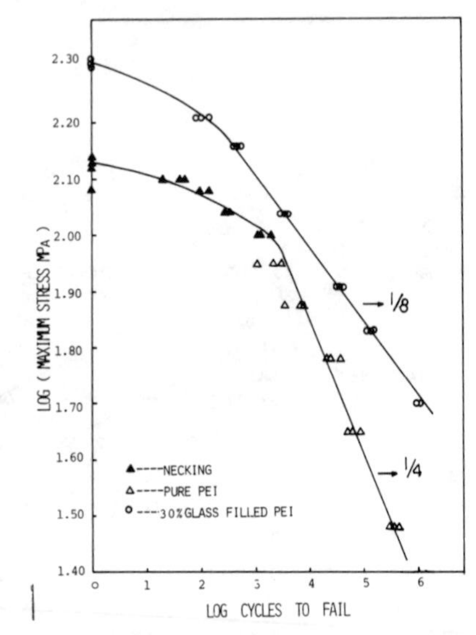

FIG. 11—S-N *data for unnotched PEI systems,* R = *0.1, 5 to 20 Hz.*

of the lifetime. The general trends of fatigue crack growth and *S-N* curves for the PSUL composites are shown in Fig. 12; the fatigue crack growth curves approach K_Q at rates on the order of 0.1 to 1 mm/cycle. The crack growth rate then decreases approximately with ${K_{max}}^m$ until a threshold, K_{th}, is reached where the crack no longer grows. The corresponding *S-N* curves show a trend with the same value of m. The higher K_Q material (graphite reinforced in this case) also has the higher UTS; this difference carries over to a similar extent to the crack growth rate at a given K_{max} as well as the lifetime, N, at a given S_{max}. Other material systems have been studied to a lesser extent, but show similar trends as long as the failure modes are consistent.

The consistency of the unnotched and crack growth data trends allows two interpretations of cause and effect. The first interpretation is illustrated in Eq 5, where the *S-N* curve is predicted to follow the power law of the crack growth data. This interpretation is rational only if the unnotched specimen fails due to fatigue crack growth. Another interpretation would be that the *S-N* curve is the basic fatigue property. It could be assumed that a region at the crack tip, a ligament of width h in Fig. 4 for example, must be cycled to failure at the local stress level to produce a crack extension of Δh. Using the fiber avoidance model in Fig. 4, Eq 3 can be generalized for fatigue to

$$K_{max}/K_Q = (2\pi l_f^*)^{1/2} S_{max}/\text{UTS} \tag{6}$$

Substituting the *S-N* curve trend into Eq 6 will yield the prediction that crack growth will follow the same exponent as the *S-N* curve, which it is observed to do. This approach is justified only if the unnotched specimen degrades generally rather than by failing from a dominant crack. Such appears to be the case with

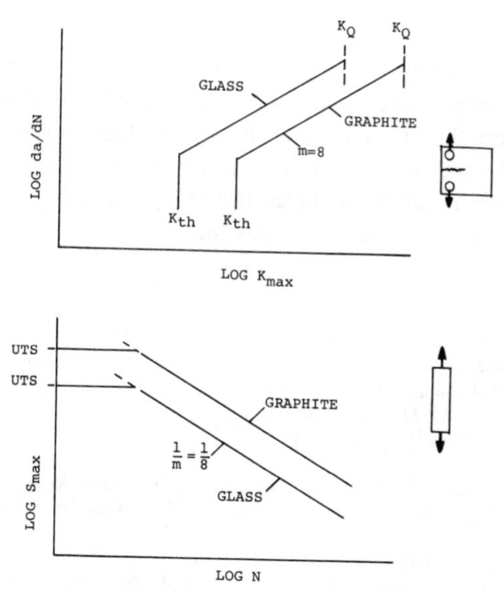

FIG. 12—*Schematic of relationship between fatigue crack growth and unnotched molded bar* S-N *behavior for injection molded composites.*

long glass fiber dominated composites, where the fatigue response is exponential rather than power law [*14*]. However, unnotched specimens of the very short fiber composites considered here fail due to fatigue crack propagation. The fundamental fatigue characteristic must be the fatigue crack growth resistance. The distinction is made by carefully observing the mechanism of breakdown in each case.

A related consideration is the role of crack initiation in the lifetime of unnotched specimens. In reinforced transparent polymers such as PSUL, it is possible to determine the approximate conditions for craze or crack initiation. Recent unpublished results for PSUL indicate that initiation either at a crack tip or in an unnotched specimen may have a similar stress or K_{max} dependence as does propagation; this does not appear to be true for some other polymers, in which case the *S-N* and crack growth data do not show the same trends. Thus, the correspondence of *S-N* and crack growth data as shown in Fig. 12 either may require that the lifetime is crack growth dominated or that initiation follows the same stress dependence as propagation. The initiation question has not yet been investigated for this class of composites.

Conclusions

Cracks in this group of very short fiber reinforced thermoplastics propagate in a matrix dominated, fiber avoidance mode. The process zone at the crack tip is of a fixed dimension determined by the microstructure, in particular the length of small regions of fiber agglomeration. The fracture toughness can be predicted from the UTS based on this model. Fatigue crack growth occurs at a power law exponent of 7 to 8, approximately twice as high as for the matrix materials. The crack growth threshold stress intensity values are a much higher fraction of the fracture toughness than for metals or amorphous thermoplastics. K_{th}-values approach those of many metals on an absolute basis. Trends of *S-N* curves for unnotched specimens can be predicted from fatigue crack growth data as long as the specimens fail by fatigue crack growth.

References

[1] Mandell, J. F., Darwish, A. Y., and McGarry, F. J. in *Test Methods and Design Allowables for Fibrous Composites, ASTM STP 734,* C. C. Chamis, Ed., American Society for Testing and Materials, Philadelphia, 1981, pp. 73–90.

[2] Mandell, J. F., Huang, D. D., and McGarry, F. J., *Polymer Composites,* Vol. 2, 1981, p. 137.

[3] Mandell, J. F., McGarry, F. J., Huang, D. D., and Li, C. G., *Polymer Composites,* Vol. 4, 1983, p. 32.

[4] Mandell, J. F., Huang, D. D., and McGarry, F. J. in *Short Fiber Reinforced Composite Materials, ASTM STP 772,* B. A. Sanders, Ed., American Society for Testing and Materials, Philadelphia, 1982, pp. 3–32.

[5] Ramsteiner, F., *Composites,* Vol. 12, 1981, p. 65.

[6] Ramsteiner, F. and Theysohn, R., *Composites,* Vol. 10, 1979, p. 111.

[7] Cottrell, A. H., *Proceedings of Royal Society of London,* Vol. A282, 1964, p. 2.

[8] "Fracture Testing of High Strength Sheet Materials," *ASTM Bulletin,* No. 243, Jan. 1960, p. 29.

[9] Hertzberg, R. W. and Manson, J. A., *Fatigue of Engineering Plastics,* Academic Press, New York, 1980, pp. 74–141.

[10] Fuchs, H. O. and Stephens, R. I., *Metal Fatigue in Engineering,* Wiley, New York, 1980, pp. 300–301.

[11] Evans, A. G., *International Journal of Fracture,* Vol. 16, 1980, p. 485.

[12] Wang, S. S. in *Composite Materials: Testing and Design (Fifth Conference), ASTM STP 674,* S. W. Tsai, Ed., American Society for Testing and Materials, Philadelphia, 1979, p. 642.

[13] Wilkins, D. J., Eisenmann, J. R., Camin, R. A., Margolis, W. S., and Benson, R. A. in *Damage in Composite Materials, ASTM STP 775,* K. L. Reifsnider, Ed., American Society for Testing and Materials, Philadelphia, 1982, pp. 168–183.

[14] Mandell, J. F. and Meier, U. in *Fatigue of Composite Materials, ASTM STP 569,* American Society for Testing and Materials, Philadelphia, 1975, p. 28.

[15] Owen, M. J. in *Short Fiber Reinforced Composite Materials, ASTM STP 772,* B. A. Sanders, Ed., American Society for Testing and Materials, Philadelphia, 1982, pp. 64–84.

[16] Lang, R. W., Manson, J. A., and Hertzberg, R. W., *Organic Coatings and Plastics Chemistry Preprints,* American Chemical Society, Washington, DC, Vol. 45, 1981, p. 778.

[17] Paris, P. and Erdogan, F., *Journal of Basic Engineering, Trans. ASME,* Vol. 85, 1963, p. 528.

[18] Pelloux, R. M. in *Proceedings,* Air Force Conference on Fatigue and Fracture of Aircraft Structures and Materials, AFFDL TR 70–144, Air Force Flight Dynamics Laboratory, Wright Patterson Airforce Base, OH, 1969, p. 409.

Golam M. Newaz [1]

Analysis of Flexural Fatigue Damage in Unidirectional Composites

REFERENCE: Newaz, G. M., "**Analysis of Flexural Fatigue Damage in Unidirectional Composites,**" *High Modulus Fiber Composites in Ground Transportation and High Volume Applications, ASTM STP 873,* D. W. Wilson, Ed., American Society for Testing and Materials, Philadelphia, 1985, pp. 51–64.

ABSTRACT: A flexural fatigue study of unidirectional E-glass/epoxy composite with 48 volume % glass fibers was undertaken (1) to understand and analyze the mechanisms of fatigue damage and (2) to evaluate the material degradation due to cyclic loading by monitoring the loss in stiffness. The fatigue tests were performed in the deflection-controlled mode with a sinusoidal waveform for a frequency of 3 Hz. The deflection-controlled testing mode was chosen to allow for decreased tendency toward hysteretic heating and to maintain controlled damage growth during fatigue cycling. Also, this testing mode minimizes the creep-fatigue interaction. Fatigue failure mechanisms in high and low deflection levels were found to be different. At high deflection levels, both matrix cracks and fiber breaks preferentially oriented at 90° to the fiber direction occurred simultaneously on the tensile side of the specimen, and out-of-plane fiber buckling was evident on the compression side, which eventually resulted in localized delamination. At low deflection levels, matrix cracking and fiber-matrix debonding were evident on the tensile surface. Fiber breakage was either not present or was very minimal. Compressive damage was not a concern at low deflection levels. The stiffness loss at high deflection cycling was rapid as opposed to a gradual decrease observed at low deflection cycling. The greater stiffness loss at the higher deflection levels was attributed primarily to glass fiber breaks.

KEY WORDS: flexural fatigue, unidirectional composite, deflection-controlled fatigue test, damage mechanisms, stiffness loss

Applications of glass fiber reinforced composite materials are receiving considerable attention in the ground transportation industry. Advances in processing, high volume production, parts consolidation, and the low cost of glass fiber reinforced composites make them competitive compared with conventional metallic materials. Corrosion resistance, lightweight, high strength-to-weight ratio, and favorable mechanical properties make them quite attractive for load-bearing structural applications. Another desirable feature associated with the use of glass

[1]Senior engineer, Fiberglas Technical Center, Owens-Corning Fiberglas Corp., Granville, OH 43023.

fiber reinforced composites is that the composite can be custom designed to provide many of the required performance properties.

Along with many different types of composites such as sheet molding compound (SMC) and woven glass fiber reinforced plastics, unidirectional composites are being considered for a variety of automotive structural applications. Fatigue behavior of the material is an important consideration in most structural applications. Repeated loading of a structural member during service is common. Understanding the fatigue behavior is therefore needed for reliable design of composite materials and the end structures. There are two motivating reasons for pursuing fatigue studies of unidirectional E-glass/epoxy composite in this investigation, namely, (*a*) to study the controlling mechanisms of fatigue damage, and (*b*) to evaluate the cyclic property degradation.

Better understanding in these areas is expected to provide guidelines in designing composites with improved fatigue resistance.

The primary objective of this study has been to understand the mechanisms of fatigue damage. At the same time, changes in stiffness were recorded to quantify the material degradation due to cyclic loading which includes various types of damage accumulation. Determining the *S-N* curve was not the primary goal of this study. These curves, while very meaningful, do not indicate the mechanical degradation due to fatigue, nor do they provide much insight into the mechanisms of fatigue damage. However, the utility of *S-N* curves for structural design cannot be denied since they provide information about the ability of the material to sustain different levels of load for different numbers of cycles.

In this investigation, flexural fatigue study of unidirectional E-glass/epoxy composites was undertaken. This area of fatigue remains mostly unexplored except for a few meaningful studies such as those by Dharan [*1*] and Fesko [*2*]. Deflection-controlled fatigue tests were run by these workers, and appropriate interpretations were given. Talreja [*3*] reports recent results obtained by Prewo and Thompson on graphite fiber reinforced glass composites. However, it is now perceived that much need exists for a better understanding of the flexural fatigue behavior of unidirectional composites. Also, no comprehensive effort has been made to understand and analyze damage mechanisms in flexural fatigue, although the issue of damage mechanisms in fatigue, in general, has received considerable attention in recent years [*3–8*]. From a physical standpoint, the nature of property degradation in composite materials due to cyclic loading can be better understood by studying the associated damage mechanisms. Considerable effort has been spent toward achieving this understanding in this investigation.

Materials and Specimens

Materials

The material used in this study was unidirectional E-glass/epoxy composite with 48% fiber by volume. An epoxy compatible Owens-Corning Type 30 E-glass roving was chosen. This roving is primarily used for filament wound

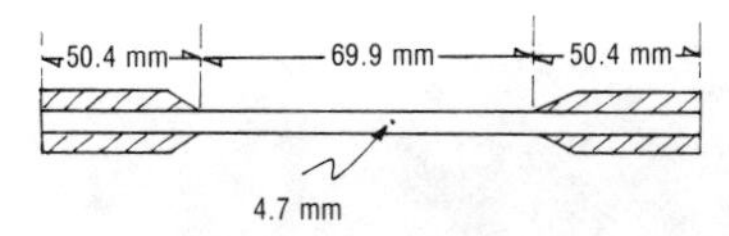

FIG. 1—*Dimensions of flexure fatigue specimen.*

applications. The epoxy chosen was Dow Epoxy Resin (DER) 331, which is commonly used for similar applications. Methylenedianiline (MDA) was used as the catalyst for curing this resin. The ratio of the resin and the catalyst mixed was 100:26.2 by weight.

The unidirectional composites were prepared in the laboratory using a small-scale filament winding machine. The glass fibers were drawn through a resin bath, a stripper die, and then were wound on flat plates. The laminates prepared were left under pressure at room temperature for 24 h. Subsequently, they were cured at 177°C for 1 h in a curing oven and cooled slowly.

Specimens

The geometry of flexural fatigue test specimens used in this study is shown in Fig. 1. The length and the width of each specimen was 170.7 and 12.7 mm, respectively. The gage length was 69.9 mm. The end tabs were 50.8 mm long with a tapering angle of 15°. The nominal thickness of the specimens was 4.7 mm. All machined specimens were subsequently sanded at the edges with care to remove machining marks or minute discontinuities. The end tab material was cross-plied glass fiber reinforced epoxy laminate and was bonded to the specimens using Hysol adhesive. The primary function of the end tabs was to protect the specimens from damage due to gripping during cyclic loading.

Experimental Approach

Flexural fatigue tests were performed using a modified flexural fatigue fixture originally designed by SATEC Systems, Inc. Figure 2 shows the fatigue fixture. When a specimen is gripped between two ends in the fixture, the loading imposed on the specimen is as shown schematically in Fig. 3. Two opposing couples are generated due to the forces of equal magnitude and acting in opposite direction resulting from the four-point loading condition in the fixture. The overall loading subjects the gage section of the specimen to pure bending.

All flexural fatigue tests were conducted at room temperature in a 89 kN Instron Servohydraulic system. The testing machine is interfaced with a PDP 11/23 minicomputer manufactured by Digital Equipment Corp. The computer was used for controlling the tests and for data acquisition purposes. Constant amplitude deflection-controlled tests were performed at a frequency level of 3 Hz for zero-tension testing mode ($R = 0$) with sinusoidal waveform. The actuator stroke was used to control the deflection in these tests. The deflection levels refer to the maximum center point deflection of the specimen. Deflection-controlled

FIG. 2—*Flexural fatigue fixture.*

testing was chosen primarily for two reasons: (1) this allows for decreased tendency toward hysteretic heating since load continues to decrease; and (2) the damage or cracks do not grow in an uncontrolled manner as would be the case in load-controlled testing. This is expected to facilitate periodic damage observation. Also, note that in this testing mode any creep-fatigue interaction will be minimized. During the deflection-controlled testing, the load drop was monitored as a function of fatigue cycles. Each test was run until either a 40 to 50% load drop from the first cycle occurred or a million cycles was reached. Tests were run at five different maximum deflection levels of 38.1, 35.6, 33.0, 28.0, and 25.4 mm. At least four specimens were tested at each level. These deflection levels correspond to initial strains on the tensile surface of the specimen as shown in Fig. 4. This deflection-strain plot was recorded using strain gage on the specimen on the tensile side. It may be noted that as damage develops during constant deflection fatigue cycling, the initial strains levels can no longer be guaranteed to remain constant. Therefore, deflection-controlled testing should not be interpreted as strain-controlled testing.

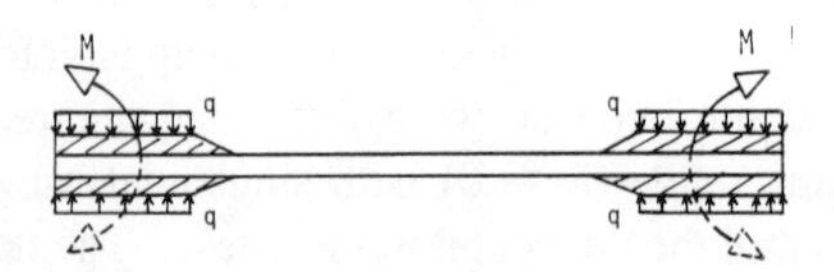

FIG. 3—*Imposed loading on fatigue specimen.*

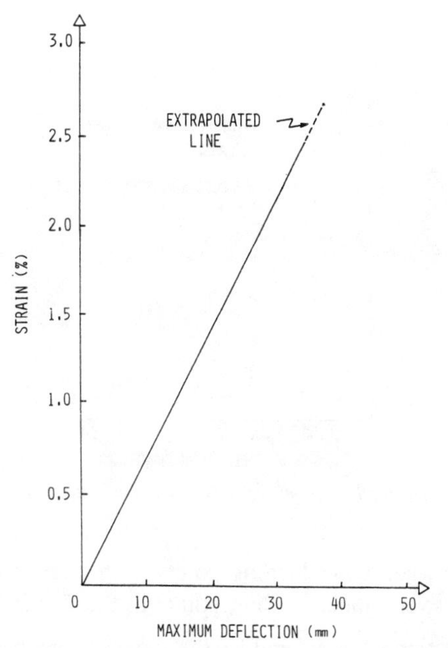

FIG. 4—*Maximum deflection versus percent strain for flexural fatigue specimen under static loading.*

In deflection-controlled testing, failure may be defined at an arbitrary percentage loss of the original load. A good discussion of this particular testing mode is presented by Hertzberg and Manson [*9*]. For this investigation, a 10% load loss criterion was selected to construct an *S-N* curve.

Results and Discussion

Stiffness Degradation Due to Fatigue Loading

As mentioned earlier, the decay in load level as a consequence of continuous cycling was monitored. An example is shown in Fig. 5. From this load versus number of cycles curve, the change in stiffness can be easily calculated as a function of number of cycles. For the loading condition illustrated in Fig. 3, the maximum deflection occurs at the midpoint of the specimen and is a function of several variables as follows:

$$\delta = kP/EI \tag{1}$$

where

P = applied load at each of the four loading points in the fixture,
EI = bending stiffness of the specimen,
k = a constant = $f(l_1, l_2)$,

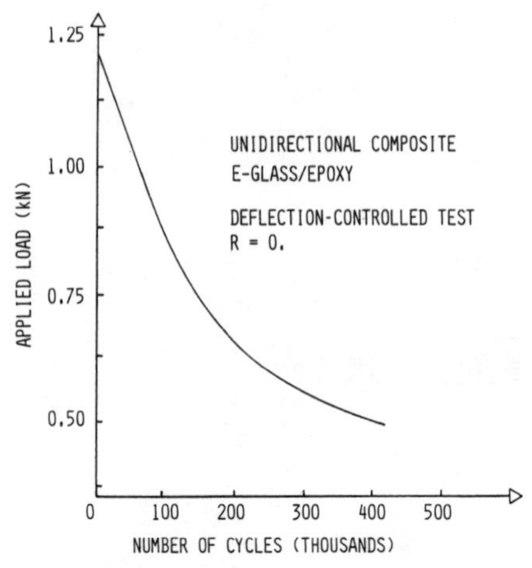

FIG. 5—*Typical load decay curve for a deflection amplitude of 35.6 mm.*

l_1 = distance between outer loading points in the flexural fixture, and
l_2 = distance between inner loading points in the flexural fixture.

For a constant deflection level δ_c, and for the same specimen geometry, Eq 1 can be rewritten as follows:

$$E = k_1 P \tag{2}$$

where

$$k_1 = k/(\delta_c I)$$

The ratio of *N*th cycle stiffness to the initial stiffness is simply given by

$$E_N/E_0 = P_N/P_0 \tag{3}$$

Since in flexural loading the specimen experiences tension and compression simultaneously and the modulus in tension and compression are not the same for unidirectional composites, the ratio calculated using Eq 3 may be termed a *pseudostiffness ratio*.

The average stiffness decay curves were constructed considering all specimens run at a particular deflection level. This is shown in Fig. 6. These curves represent an overall stiffness degradation behavior of the composite as a function of number of cycles. From these results it is observed that after about a thousand cycles the stiffness degradation is rapid for the 38.1-mm deflection case. For the 35.6-mm deflection level, the stiffness loss closely follows the previous case except that an increase in life of the member is observed. As the deflection level decreases, life is increased and stiffness loss becomes relatively more gradual. For the lowest deflection level of 25.4 mm, barely a 10% loss in original stiffness is observed up to 1 million cycles.

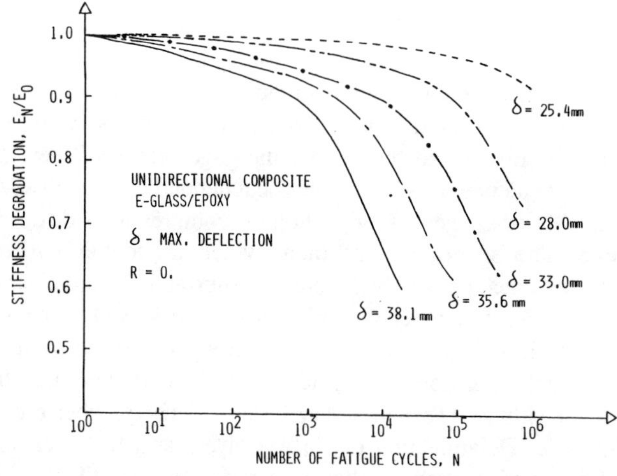

FIG. 6—*Stiffness degradation versus flexural fatigue cycles for deflection controlled testing.*

The stiffness degradation is a consequence of multiple cracking that occurs in composite materials. This behavior is significantly different than what is observed in homogeneous materials, especially in metals. In metals, fatigue loading may result in the initiation of a single crack which will continue to grow until failure occurs. During the crack growth process, the material stiffness remains unchanged. For composite materials, it is quite clearly recognized that the stiffness degradation process is unique and occurs due to multiple cracking in the laminate. The stiffness loss is essentially a macroscopic measure of the state of overall damage in the composite. In the next section, the damage mechanisms are discussed in detail.

Damage Mechanisms

In zero-tension flexural loading, one side of the specimen above the neutral axis is always subjected to tension and the other portion is always subjected to compression. Since only pure bending stresses were generated, shear stresses through the thickness were absent. Therefore, damage accumulation may be considered to be due to axial tension and compression. However, it is recognized that shear stresses could come into play as a consequence of damage development at localized regions as cycling is continued.

During fatigue cycling, within the first few cycles, the tapered portion of the end tabs separated from the specimen and acted as flexible layers without providing any constraints to the bending of the specimen. To eliminate the possibility of examining any extraneous damage growth which may have resulted from gripping areas, the central portion of the specimens was examined. It may be mentioned that the specimen was designed to avoid grip damage. Only minimal damage was observed in some cases at the grip area.

At a high deflection level of 38.1 mm, a specimen was cycled for only 2 cycles and was taken out for inspection using an optical microscope. On the tensile side, numerous microcracks oriented at 90° to the maximum stress direction could be seen as shown in Fig. 7. Both matrix cracks and fiber breaks were observed and were found to be distributed throughout the gage section. Where fiber breaks occurred, some evidence of splitting and local debonding was observed adjacent to these breaks. No damage was evident on the compression side of the specimen at this stage. The specimen was then cycled up to 500 cycles and then reexamined. The crack density had increased substantially and clear evidence of more splitting cracks with much larger lengths was observed. Some surface layer delamination of glass fiber strands was also observed. On the compression side, out-of-plane buckling of fibers was evident. This occurred because there was no constraint for buckling. There was very little evidence of fiber breakage on the compression side. Delamination of surface layers at a few locations was also observed. Continued cycling resulted in a rapid drop in stiffness. This can be related to the glass fiber breaks on the tensile side, which were extensive. Qualitatively, the damage on the tensile side was much more extensive than what was observed on the compressive surface.

At a low deflection level of 25.4 mm, periodic checks were made to observe microcracks in the specimen. A check at 20 000 cycles showed no microcracks. At the next observation at 50 000 cycles, matrix microcrack normal to the tensile stress direction could be observed (Fig. 8). The microcrack density was much lower compared to the first observable microcracks very early in the life cycle for the 38.1-mm specimen. The microcracks generated due to low deflection cycling were also found to be sparsely distributed. After continued loading to 200 000 cycles, some splitting was evident. At about half a million total number of cycles, numerous splitting cracks and strand delamination was observed. However, there was very little evidence of glass fiber breaks resulting from transverse cracking on the tensile surface compared to the 38.1-mm deflection specimens. On the compression side, at this low deflection level, no damage was observed until nearly 1 million cycles. Very little delamination due to the tendency of out-of-plane buckling of fiber strands could be observed at this stage. This occurred near the edges of the specimen. Compared to the compression damage in 38.1-mm deflection specimens, this damage was negligible.

The major differences between microcrack formation at high and low deflection levels are as follows:

(*a*) both matrix cracks and glass fiber breaks initiate simultaneously at high deflection levels, and

(*b*) only matrix cracks occur early at low deflection levels.

An analogy of these damage processes can be drawn from metal fatigue. At high stress levels, the fatigue crack initiation in metals is controlled by the ductility of the material. And at low stress levels, it is controlled by the strength of the material. The fiber is relatively more brittle compared with the resin and is expected to fail earlier at high stress levels. The simultaneous formation of matrix cracks is not very well understood. However, it can be proposed that

FIG. 7 — (a) *Matrix and fiber damage on tensile surface;* (b) *fracture of glass fibers. The specimen was fatigued for 2 cycles at a deflection amplitude of 38.1 mm.*

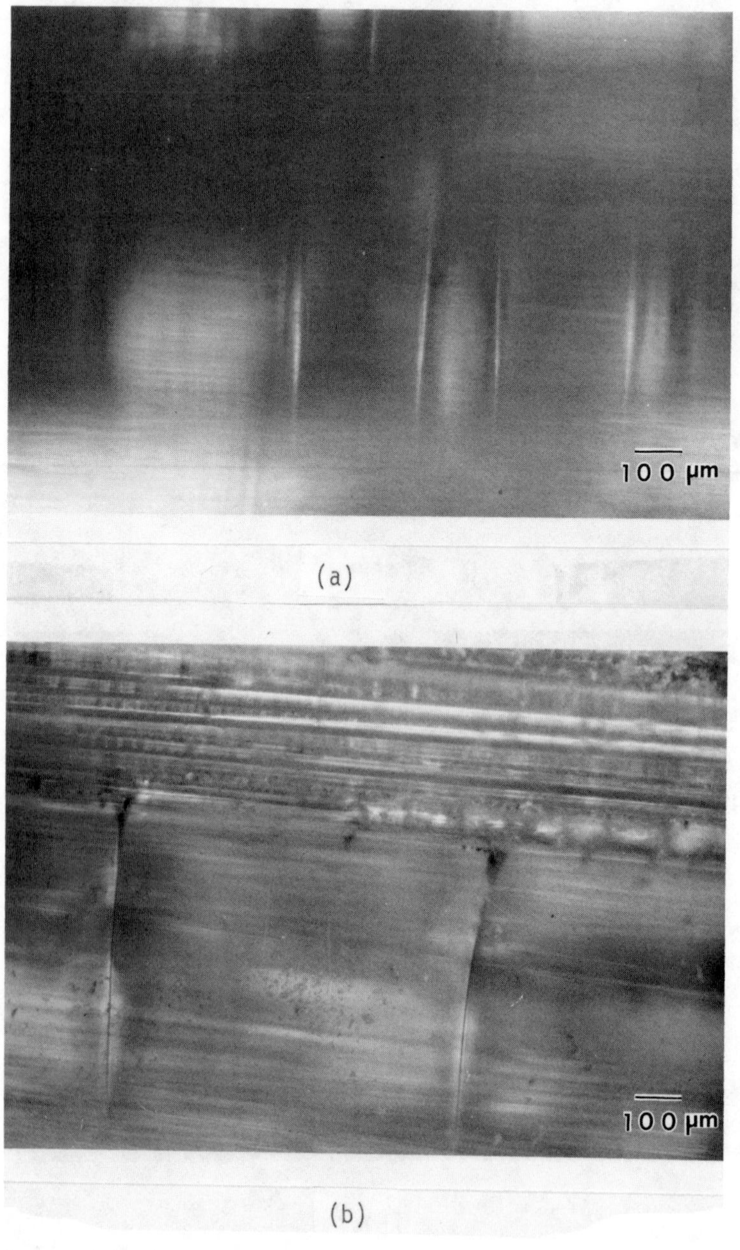

FIG. 8—(a) *Subsurface matrix cracks;* (b) *surface matrix cracks. The specimen was subjected to 50 000 cycles at a deflection amplitude of 25.4 mm.*

constraints due to the fibers and the local curing characteristics may strongly influence the resin fracture. At low stress levels, the glass fibers are expected to perform better in fatigue compared to the more ductile phase, the resin. This was observed as mentioned earlier in (*b*).

Damage investigation was also conducted with specimens at intermediate deflection levels. The nature of microcracks and progressive damage development could be characterized as intermediate with respect to the damage processes caused by low and high deflection fatigue cycles.

Some evidence of splitting cracks and fiber breaks was found near the end tabs. The splitting cracks were possibly due to the Poisson's effect as load was imposed on the specimen. The fiber breaks may be related to the unwanted stress concentrations at the end tabs. However, compared to the major damage at the gage section, damage at the grip ends was minimal.

One of the most interesting findings of this study has been that many of the microcracks normal to the principal tensile stress direction were found to have originated at subsurface locations rather than at the surface. Subsequent scanning electron microscope (SEM) study with a virgin specimen shows that fibers are embedded at various depths below the surface layer of the resin (Fig. 9). To clearly distinguish the surface layer, an alumina-filled layer was bonded to the cross-sectional piece of the specimen for SEM observation. It is possible that many of the microcracks initiated at fiber-matrix interphase locations rather than on the surface of the specimen. The polymeric material at the interphase often does not seem to have the properties of the bulk resin. Crack initiation at fiber-matrix interphase is very common in glass fiber reinforced composites.

Damage Map and S-N *Curve*

We have discussed both the damage development and stiffness degradation in an E-glass/epoxy unidirectional composite. It may be appropriate to consider damage mapping to illustrate the stiffness loss due to the particular state of damage during fatigue loading. The two extreme deflection cases of 38.1 and 25.4 mm have been considered. The fatigue damage map is shown in Fig. 10. This figure demonstrates the overall composite damage process and simultaneous property loss as a function of number of fatigue cycles.

A typical *S-N* curve is plotted using a 10% load loss criterion. It may be noted that any arbitrary load loss criterion can be selected, as shown in Fig. 11, to construct an *S-N* curve provided enough fatigue data are generated. For the flexural loading using the particular test fixture, the maximum tensile stress at the outer surface at a load level of P' is given by

$$\sigma_{\max} = 3\frac{P' l_1}{b \times h^2} \tag{4}$$

where

P' = load,

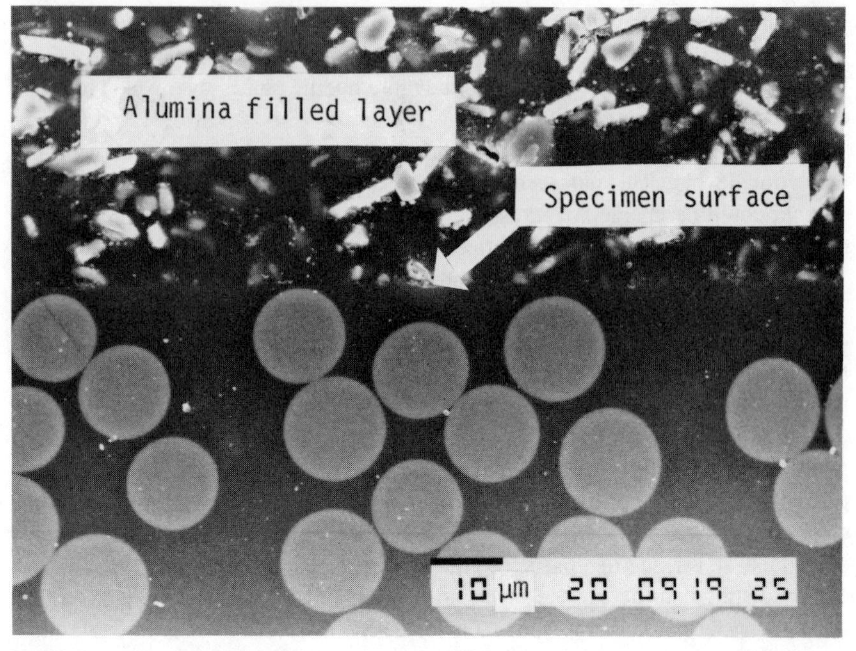

FIG. 9—*SEM photomicrograph showing variation in resin layer depth from specimen surface.*

b = specimen width, and
h = specimen thickness.

A 10% load decay corresponds to $P' = 0.9P_0$, where P_0 is the maximum initial load during fatigue cycling. The load-decay curves are used to determine the fatigue life at this load level. The stress calculated using Eq 4 was plotted against fatigue life of the specimen as shown in Fig. 11. This curve relates the number of fatigue cycles of constant deflection survived by the specimen when the maximum stress level had decayed to the particular stress level. Compared with conventional *S-N* curves, it may be noted that the stress value plotted on the ordinate is not the stress amplitude.

Summary

1. Fatigue damage in a unidirectional composite in pure flexural loading is controlled by matrix cracks and glass fiber breaks, as well as by splitting on the tensile side and by out-of-plane fiber buckling which results in delamination on the compressive surface.

2. At high deflection levels, both matrix microcracks and glass fiber breaks occur very early in life. These cracks are preferentially oriented at 90° to the tensile stress direction. Continued cycling results in extensive damage in the form of splitting, glass fiber breaks, matrix cracking, and glass strand delamination.

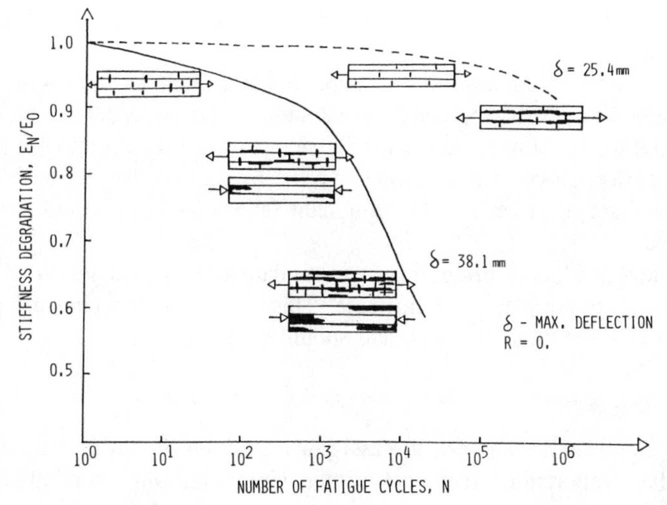

FIG. 10—*Damage map illustrating damage development on the tensile and compressive sides and corresponding stiffness degradation.*

At low deflection levels, only matrix microcracks are formed early in life followed by a few glass fiber breaks. Orientation of these cracks is 90° to the tensile stress direction. Continued cycling results in quite extensive splitting cracks.

3. The rapid stiffness loss in specimens tested at high deflection levels is primarily due to glass fiber breaks. The stiffness loss for low deflection cycling at 25.4 mm is very gradual and shows less than 10% drop during fatigue cycling

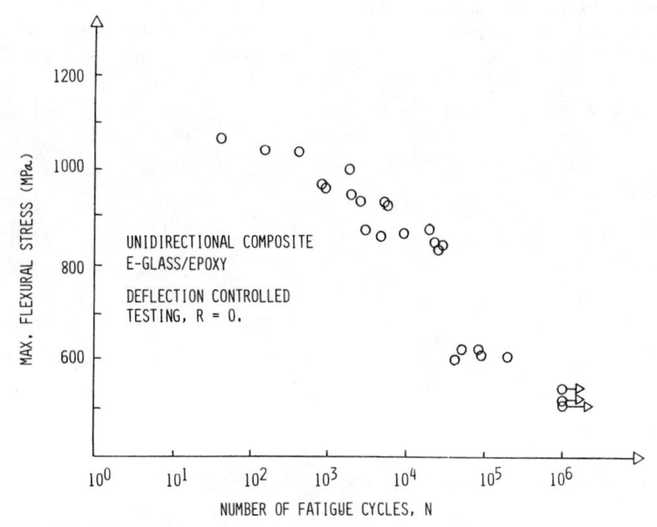

FIG. 11—*Flexural stress versus fatigue cycles using 10% load loss criterion.*

up to 1 million cycles. This is attributed to the fact that glass fiber breaks were minimal.

4. Compressive damage is only a concern at high deflection testing. Even at high deflection levels, the compressive damage did not appear to be extensive compared to the damage on the tensile side. Progressive decrease in load level during deflection-controlled testing may be responsible for not enabling compressive damage to become the dominant failure mode as would normally be expected.

5. The individual contribution of various damage modes to the overall stiffness degradation is not clearly identifiable. This would be an interesting and challenging area for future research and should be addressed.

Acknowledgment

The author acknowledges the assistance of John Sutton of Owens-Corning Fiberglas Corporation, Technical Center, Granville, Ohio, for fatigue testing.

References

[1] Dharan, C. K. H., *Journal of Materials Science,* Vol. 10, 1975, pp. 1665–1670.
[2] Fesko, D. G., *Polymer Engineering and Science,* April 1977, Vol. 17, No. 4, pp. 242–245.
[3] Talreja, R. in *Fatigue and Creep of Composite Materials,* Proceedings of the 3rd Riso International Symposium on Metallurgy and Materials Science, L. Lilholt and R. Talreja, Eds., Riso National Laboratory, Roskilde, Denmark, Sept. 1982, pp. 137–153.
[4] Talreja, R., *Proceedings of Royal Society of London,* Vol. A378, 1981, pp. 461–475.
[5] Reifsnider, K. L. and Talug, A., *International Journal of Fatigue,* Jan. 1980, pp. 3–11.
[6] Reifsnider, K. L., Henneke, E. G., Stinchcomb, W. W., and Duke, J. C. in *Proceedings,* Symposium on Mechanics of Composite Materials, Blacksburg, VA, International Union of Theoretical and Applied Mechanics, Z. Hashin and C. T. Herakovich, Eds., Pergamon Press, New York, Aug. 1982, pp. 399–420.
[7] Stinchcomb, W. W. and Reifsnider, K. L. in *Fatigue Mechanisms, ASTM STP 675,* American Society for Testing and Materials, Philadelphia, 1979, pp. 762–787.
[8] Reifsnider, K. L., *International Journal of Fracture,* Vol. 16, 1980, pp. 563–585.
[9] Hertzberg, R. W. and Manson, J. A., *Fatigue of Engineering Plastics,* Academic Press, New York, 1980.

Anh-Dung Ngo,[1] *Suong V. Hoa,*[1] *and Thiagas S. Sankar*[1]

Axial Fatigue of SMC-R65 Sheet Molding Compound in Liquid Environments

REFERENCE: Ngo, A.-D., Hoa, S. V., and Sankar, T. S., **"Axial Fatigue of SMC-R65 Sheet Molding Compound in Liquid Environments,"** *High Modulus Fiber Composites in Ground Transportation and High Volume Applications, ASTM STP 873,* D. W. Wilson, Ed., American Society for Testing and Materials, Philadelphia, 1985, pp. 65–72.

ABSTRACT: The fatigue behavior of an SMC-R65 sheet molding compound under axial fatigue loading and in the presence of water and *iso*octane is investigated. There is about a decade decrease in fatigue lives for specimens tested in water as compared to those tested in air. The fatigue lives of specimens tested in *iso*octane range from those in air to those in water. SMC-R65 specimens immersed in water show clear absorption boundaries.

KEY WORDS: sheet molding compound, axial fatigue, liquid environment

In an effort to investigate the effect of liquid environments on the fatigue behavior of sheet molding compounds (SMCs), the fatigue crack propagation of SMC-R30 and SMC-R65 in air, water, and *iso*octane was studied [*1*]. Subsequently, the effect of water and *iso*octane absorption on the flexural fatigue strength of SMC-R30 was investigated [*2*]. It was found in Ref *1* that water absorbs into SMCs more than *iso*octane and the rate of crack growth in water accelerates faster than the rate of crack growth in *iso*octane. It was found in Ref *2* that the absorption of water and *iso*octane also decreases the fatigue lives of SMC-R30 as compared to those of dry specimens. In the continuing effort to understand the effect of liquid absorption on the fatigue behavior of SMCs, the axial fatigue behavior of SMC-R65 specimens that were previously immersed in water and *iso*octane was investigated.

Experiments

The material studied was SMC-R65 sheet molding compound supplied by Budd Co. under the code number DSM-750. The material composition is shown

[1]Graduate student, associate professor, and professor, respectively, Department of Mechanical Engineering, Concordia University, Montreal, P.Q., Canada.

TABLE 1—*Material specifications of SMC-R65.*

Material[a]	Specification
Polyester resin	35.8 weight %
Calcium carbonate filler	0.35 weight %
E-glass	62.5 weight %
Balance (thickness, internal release, and catalyst)	1.35 weight %
Tensile strength	172 to 210 MPa[b]
	150 MPa[c]

[a]Supplier, Budd-Code No. DSM-750.
[b]Supplier's specifications.
[c]Found in our laboratory.

in Table 1. The fatigue specimen size and shape is shown in Fig. 1. This shape and size was chosen following the successful results obtained from similar specimens used in Ref *4*. The liquids used were water and reagent grade *iso*octane.

SMC-R65 specimens were immersed unstressed in the liquids for a period of 40 days at room temperature. The average percent weight uptake of liquids in the different specimens is 2.0% for water and 1.24% for *iso*octane. It was noted that specimens immersed in water show visible boundaries between the absorbed region and the center of the specimens (Fig. 2). Similar lines were not observed in SMC-R65 specimens immersed in *iso*octane and SMC-R30 specimens immersed in water and *iso*octane. Observation under the microscope shows that the resin within this absorbed region has swollen, and many more fibers protruded out of the specimen edge as compared to the dry specimens (unimmersed).

Fatigue tests were performed on an axial MTS servohydraulic machine using load control mode and haversine waveform. The already immersed specimens were completely immersed in the liquid throughout the fatigue test (Fig. 3).

The load was varied from 0 to tension stress with maximum stress of 80, 70, 60, 50, and 40% of the tensile strength of the material (150 MPa). The frequency of testing was 5 Hz.

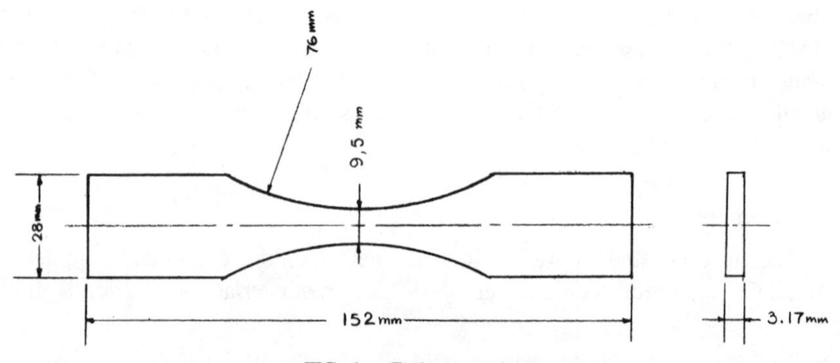

FIG. 1—*Fatigue specimen.*

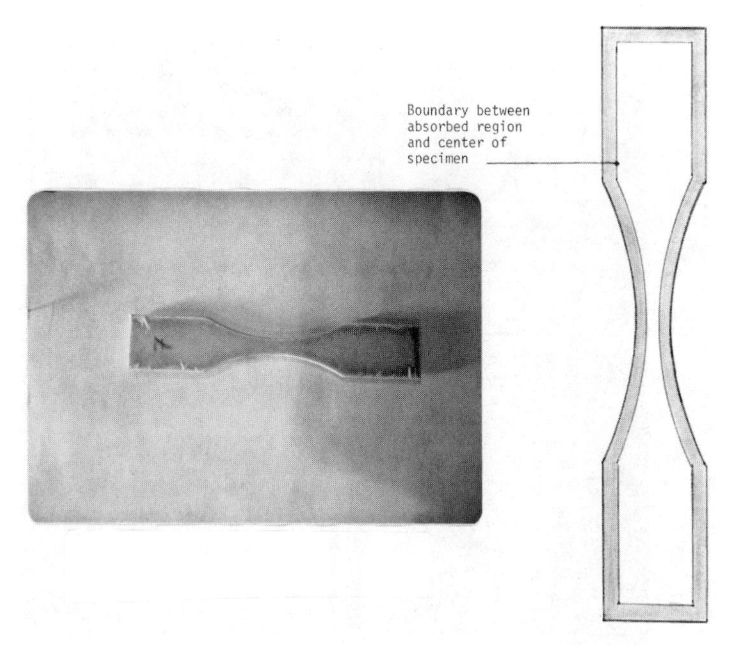

FIG. 2—*Appearance of absorbed specimen.*

Results

The fatigue life was defined as the number of cycles to specimen fracture. The fatigue lives are shown in Table 2. The variation of the fatigue lives versus the maximum stress is shown in Fig. 4.

It can be seen that the absorption of water decreases the fatigue lives by about a decade from the fatigue lives of specimens tested in air. The results obtained from specimens tested in *iso*octane show more scatter than those tested in air or in water.

Observation of the specimen edge shows that significant cracking and delamination have occurred at the edge of specimens immersed in water. By painting the surface of the specimen using india ink, it was possible to detect the crack lines (Fig. 5). Taking the specimen out of water for a few hours made the absorbed layer change color from bluishness to whitishness, and significant delamination was observed without any extra external stress. By putting the specimen back into water, the absorbed layer changed color back to bluishness again within a short time (much shorter than the time for the layer to reach its width dimension). Microhardness measurements on the surface of the specimen showed a decrease of the surface hardness on the water-immersed specimen as compared to the dry specimen. However, there was no significant variation in the surface hardness from the edge to the center of the specimen.

Figure 6 shows the broken specimens. Specimen 6 was immersed in water, Specimen 12 was immersed in *iso*octane, and Specimen 16 was tested in air

FIG. 3—*Fatigue test setup.*

(Table 2). It can be seen that delamination is a major mode of failure in the specimens immersed in water. Specimen 6 shows that delamination had split the thickness of the specimen and that one section had fallen off. The specimen that was immersed in *iso*octane also shows delamination but to a lesser extent. The failure mode of the specimen tested in air can be seen to be due to fiber pullout rather than to delamination.

Discussion

The appearance of the absorption line in water-immersed specimens is an interesting phenomenon. This phenomenon was not observed in the case of SMC-R30 specimens immersed in water [*3*]. The appearance of this line can be due to many parameters:

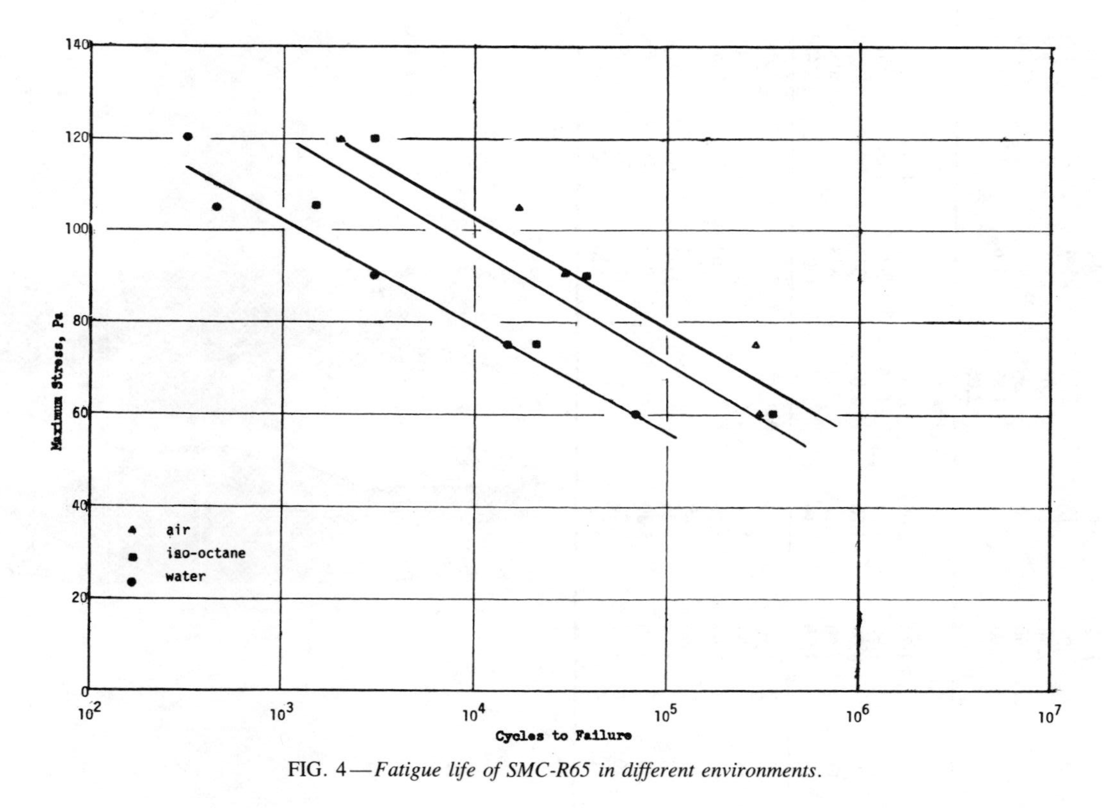

FIG. 4—*Fatigue life of SMC-R65 in different environments.*

TABLE 2—*Fatigue lives of SMC-R65 in different fluids.*

Specimen	Minimum Width, mm	Thickness, mm	Percent Weight Uptake	Load, *N*	Maximum Stress, MPa	*N*, Cycle
			WATER			
1	7.92	3.14	...	2984	120	320
3	7.62	3.18	1.99	2544	105	460
5	7.94	3.22	2.03	2301	90	3000
6	7.90	3.20	1.99	1896	75	15210
2	7.96	3.30	1.87	1576	60	71190
7			2.06			
			*ISO*OCTANE			
9	7.86	3.06	1.32	2886	120	2940
12	7.90	3.10	1.45	2571	105	1510
8	8.00	3.26	1.07	2347	90	37790
11	7.66	3.20	...	1838	75	20520
10	8.00	3.14	1.11	1507	60	366470
			AIR			
15	7.92	3.10	...	2946	120	2000
16	7.92	3.10	...	2762	113	10670
17	7.92	3.26	...	2271	105	1750
18	8.00	3.26	...	2349	90	29290
19	7.90	3.30	...	1955	75	266400
20	7.90	3.28	...	1555	60	3.5×10^6
21	7.90	3.22	...	1526	60	303780

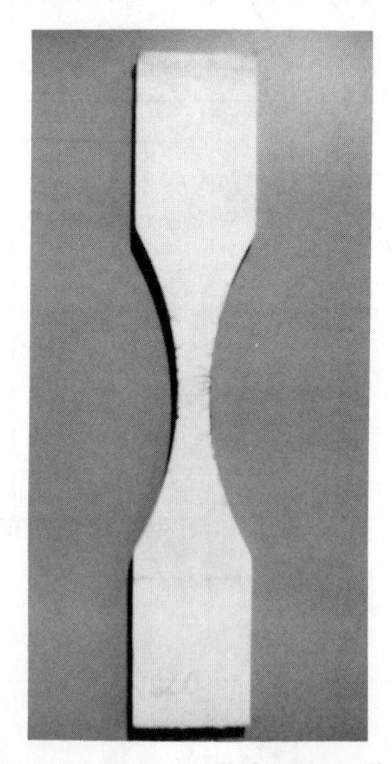

FIG. 5—*Crack lines on SMC specimens (in air).*

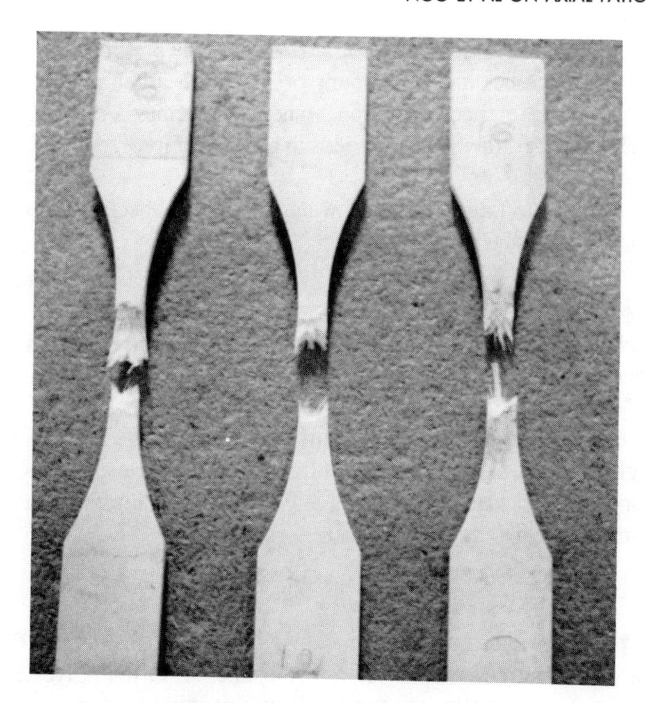

FIG. 6—*Broken samples:* (left) *sample 6, water;* (center) *sample 12,* isooctane; (right) *sample 16, air.*

Edge Cracking Due to Machining

By immersing a specimen with three edges machined and one edge as molded (unmachined), it was found that the absorption line appeared on the machined edges while no absorption line appeared on the unmachined edge. Also, by painting a machined edge with polyurethane, no absorption line was observed. These evidences indicate that cracking could have been created due to machining. The presence of these cracks allows the rapid permeation of water by capillary action. However, edge cracking due to machining cannot be the only factor responsible for the appearance in the absorption line since machined specimens of SMC-R30 did not show these absorption lines [*3*].

Interaction Between Polyester Resin and Water

Even though SMC-R65 contains more fibers than SMC-R30 (62.5 weight % for SMC-R65 as compared to 30 weight % for SMC-R30 [*1*]), SMC-R65 contains more resin than SMC-R30 (35.8 weight % for SMC-R65 and 27 weight % for SMC-R30). This is because SMC-R30 contains more calcium carbonate filler (41 weight % for SMC-R30 and 0.35 weight % for SMC-R65). However, this difference in polyester resin content does not make a difference in the percent weight uptake in the two SMCs. Reference *1* showed that at 80 days, the percent weight uptake for SMC-R30 is 2.17 and for SMC-R65 is 2.28. The appearance of the absorption line therefore cannot be attributed to the presence of more resin. The significant difference in the amount of calcium carbonate filler may be responsible for the difference in the appearance of specimens of SMC-R30 and SMC-R65. However, further analysis needs to be done before a definite statement can be made.

The reduction in the fatigue lives of water-immersed specimens can be also attributed to two parameters:

Presence of Cracks from Machining

Since machining was shown to contribute to the appearance of the absorption line and many cracks were observed to appear on the immersed machined edge, cracks due to machining can contribute to the reduction in fatigue lives. The quantitative determination of this contribution remains to be done, however, in order to single out the effect of water absorption alone on the strength of SMCs; it is necessary to avoid machining the edge either by molding the specimens or by painting the machined edge with resin.

Swelling Due to Absorption and Loosening the Interface

The absorption of water into SMC may swell the resin, may attack the fiber-resin interface, and may eventually loosen the fiber from the polyester. These factors are all possible and are the topics of current investigation.

Conclusion

The effect of water and *iso*octane on the axial fatigue of SMC-R65 was investigated. Water decreases the fatigue lives of SMC-R65 by about a decade from that in air. *Iso*octane also reduces the fatigue lives but in a more irregular manner. Absorption of water clearly shows a boundary layer and creates delamination at the edge of the specimen. The failure mode of specimens immersed in *iso*octane was partially due to delamination whereas that of specimens tested in air was due to fiber pullout and fiber breakage.

Acknowledgment

The financial assistance from the Natural Science and Engineering Research Council of Canada through grant no. A0413 is greatly appreciated.

References

[*1*] Hoa, S. V., Ngo, A. D., and Sankar, T. S., "Fatigue Crack Propagation of Sheet Molding Compounds in Various Environments," *Polymer Composites,* Vol. 2, No. 4, Oct. 1981.

[*2*] Hoa, S. V., Ngo, A. D., and Sankar, T. S., "Effect of Water and Isooctane Absorption on the Flexural Fatigue Strength of a Sheet Molding Compound," *Polymer Composites,* Vol. 3, No. 1, Jan. 1982.

[*3*] Hoa, S. V. and Ouellette, P., "Liquid Absorption of a Sheet Molding Compound," *Polymer Composites,* Vol. 2, No. 4, Oct. 1981.

[*4*] Owen, M. J., *Composites,* Vol. 1, No. 6, Dec. 1970, pp. 346–355.

Dale W. Wilson [1]

Characterization of Bolted Joint Behavior in SMC-R50

REFERENCE: Wilson, D. W., **"Characterization of Bolted Joint Behavior in SMC-R50,"** *High Modulus Fiber Composites in Ground Transportation and High Volume Applications, ASTM STP 873,* D. W. Wilson, Ed., American Society for Testing and Materials, Philadelphia, 1985, pp. 73–85.

ABSTRACT: Results are reported for an experimental investigation into the fundamental characteristics of bolted joint behavior in SMC-R50. Single fastener, double lap test specimens were employed to evaluate strength and failure mode behavior as a function of clamping pressure, fastener size, fastener edge distance, and fastener half spacing. Although three failure modes were observed—bearing, cleavage, and net tension—the net tension failure mode dominated through a broad range of geometries. This contributed to an observed difference in strength trends observed for the SMC-R50 material systems as compared to continuous fiber laminate systems. The implications of the empirical strength and failure mode findings on design and analysis of mechanical joints in SMC-R50 are discussed.

KEY WORDS: composite material, sheet molding compound (SMC), bolted joint

While the attachment of sheet molding compound (SMC) components through adhesive bonding is preferred, the use of mechanical fasteners is unavoidable in some situations. The use of mechanical fasteners requires the removal of material from a component and the insertion of fasteners (usually metallic) with significantly different elastic properties from the SMC. The resultant stress concentration reduces the load-carrying capacity of the component and can be a source of catastrophic failure.

The behavior of composite bolted joints for continuous fiber laminated systems has been studied extensively [*1–10*]. Early efforts were directed at empirical characterization of bolted joint behavior for a given material system and class of laminates. Later efforts began focusing on analytical models in attempts to reduce the amount of empirical data required to design a bolted joint [*11–16*]. After two decades of study, work continues in an attempt to improve the analytical methods available.

[1]Associate scientist, Center for Composite Materials, University of Delaware, Newark, DE 19716.

The problem of jointing in composites is difficult, because every change in geometry, material configuration, fastener size, fastener torque, and area of constraint changes the failure characteristics (strength and mode). With a continuous fiber system, the material configuration is determinant, while SMC materials present the new challenge of an indeterminant material configuration. The fiber orientation state and distribution in SMC varies from point to point and the behavior of a bolted joint depends upon the properties of the material in a critical region surrounding the fastener cutout. Macroscopically, SMC material is designed to be nearly isotropic in plane. The critical question is "Can an SMC bolted joint be designed on the basis of homogenous, isotropic, or orthotropic assumptions or is a more realistic approach needed that accounts for the fiber orientation distribution of the material?"

In this paper, results from the first phase of a three-phase project are reported. The first phase is the characterization of failure mechanisms for single fastener, double lap bolted joints in SMC materials. While the use of a double lap configuration in practice is unlikely for SMC components, it is the most basic system to analyze and interpret. This allowed the development of a fundamental understanding of the material response in a bolted joint loading configuration.

Experimental Method

The objective of this research was the characterization of bolted joint strength and failure mode behavior for SMC-R50. The methodology employed was the systematic study of the basic material properties followed by a series of studies on bolted joint behavior while varying geometric and clamping pressure parameters. Four parameters were varied independently: fastener size, edge distance, half spacing, and clamping pressure.

Material and Specimen Fabrication

The SMC material (SMC-R50) was obtained from Owens-Corning Fiberglas as cured panels 30 by 53 by 0.25 cm in dimension. The formulation and processing used to make these panels was similar to that described in Denton's monograph [*17*]. The only recorded difference was that the panels were larger and required the SMC charge to be changed in order to keep the mold coverage the same (74%). Basically, the SMC-R50 material consists of a polyester resin, 25-mm-long glass roving, calcium carbonate filler, resin inhibitor and initiator, alkaline earth oxide, and mold release. The glass roving reinforcement comprises 50 weight % of the composition making this a structural grade of SMC. For exact details of the formulation and processing conditions the reader is referred to Denton's work [*17*].

Although the processing conditions were designed to minimize flow, hence inhibit orientation, some orientation occurs during compounding. In the panels used in this program, the slight fiber alignment was expected to be in the 53 cm direction of the panel.

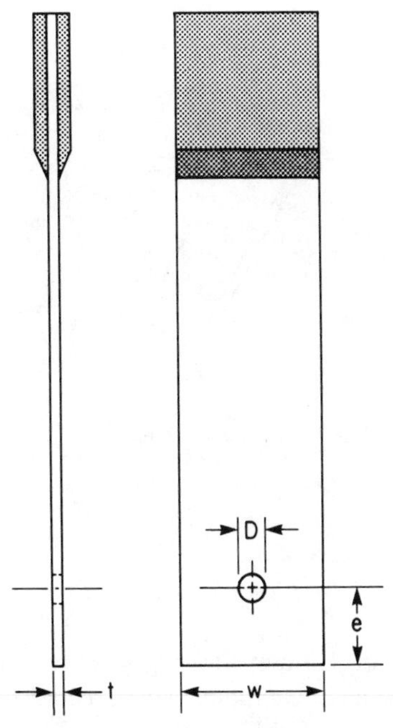

FIG. 1—*Specimen geometry.*

Single fastener, double lap bolted joint test specimens were fabricated from the SMC panels to the configuration shown in Fig. 1. All cutting and drilling was performed using diamond-impregnated cutting tools to minimize machining damage. Each specimen was inspected both visually and ultrasonically to detect any flaws or machining-induced damage, and width, thickness, hole diameter, and hole location measurements were recorded for each specimen.

Similar care was taken to fabricate the 2.5 by 23-cm tension specimens and the 1.3 by 14-cm Illinois Institute of Technology Research Institute (ITTRI) compression specimens. Specimens were fabricated from each of the two principal directions of the panel to assess the degree of orthotropy in elastic properties and strength.

Test Method

Standard test procedures were employed in the determination of tension and compression properties. The tension tests were performed in accordance with the specifications outlined in ASTM Method for Tensile Properties of Fiber-Resin Composites [D 3039-76 (1982)]. Longitudinal and transverse strains were measured by a 0/90 strain rosette (micromeasurements EA-06-125TQ-350) and used to determine the modulus and Poisson's ratio. The IITRI compression test method, ASTM Method for Comprehensive Properties of Unidirectional or Crossply Fiber-Resin Composites [D 3410-75 (1982)], was employed to measure compression properties. The specimens were 1.9 wide by 14 cm long with an

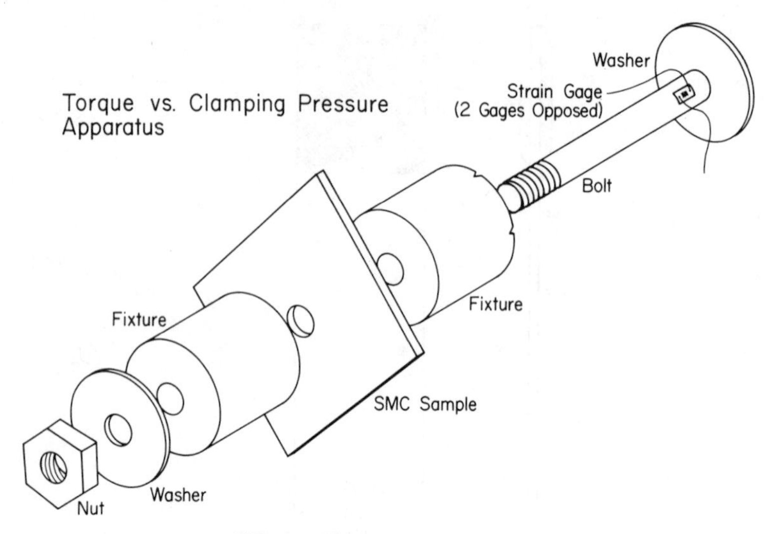

FIG. 2—*Clamping pressure fixture.*

unsupported gage length of 1.3 cm. Strain gages were mounted on opposing surfaces of the test section and longitudinal strain recorded. Longitudinal modulus was determined by averaging the two strains.

Two special fixtures were designed for the bolted joint testing and determination of clamping pressure versus torque behavior. The measurement of clamping pressure as a function of torque was performed using the fixture shown in Fig. 2. Since the relationship between torque and clamping pressure is a function of fastener diameter, components were fabricated for each of the three fastener diameters used in this test program. The test bolts were instrumented with two strain gages; the average of their strain output was used in determining the axial stress in the bolts at a specific torque level. The tensile modulus of each test bolt was determined which, together with the strain data, allowed the calculation of the effective clamping pressure as a function of torque.

The bolted joint tests were performed using the fixture shown in Fig. 3. The SMC specimens were bolted into the fixture using high strength steel fasteners with washers having an outside diameter of 2.5 times that of the fastener. Fastener-to-washer and fastener-to-hole fit were carefully monitored, since both of these factors significantly affect bolted joint strength.

The bolted joint specimens were loaded monotonically to failure in tension and the load-displacement data recorded. The failure load for the specimen was considered the highest load attained before catastrophic failure. Failure strengths were determined as the far field stress in the specimen at failure.

Results and Discussion

The results are discussed in the sequence of their execution, mechanical properties, clamping pressure-torque relationships, and bolted joint behavior.

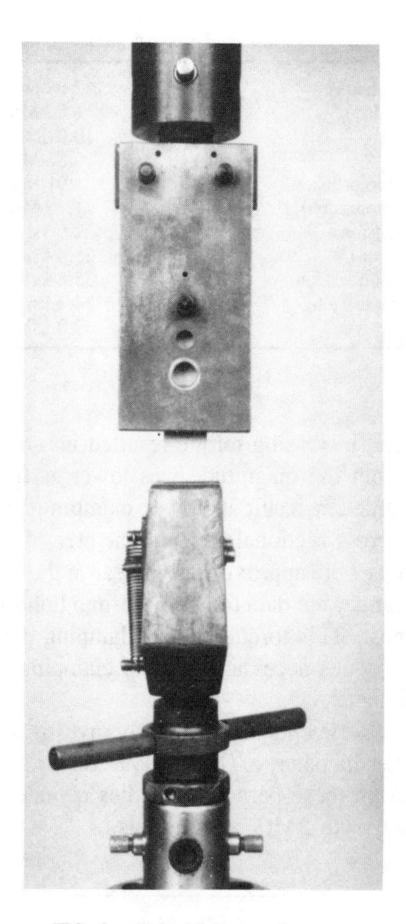

FIG. 3—*Bolted joint test fixture.*

The tension and compression strength and modulus were measured in the two principal material directions. The results were compared with those found by Denton [*17*] for similar SMC-R50 material and were found to agree within the allowed variance. Values measured for the tension and compression properties are given in Table 1, along with shear properties from Ref *17*. It is clear that a slight amount of fiber orientation has produced some orthotropy in properties. One important feature of the materials behavior is illustrated in Fig. 4, its nonlinear stress-strain response. This was observed for both tension and compression loading, and the reported stiffnesses are the initial tangent modulus. Nonlinear stress-strain behavior could complicate further the development of an analytical model for failure analysis.

Clamping pressure generated as a function of torque was measured for three bolt diameters, 6.35, 9.53, and 12.70 mm. The relationship between torque and clamping pressure for the three fastener sizes is clearly seen in Fig. 5. Indepen-

TABLE 1—*Material property data for SMC-R50.*

Longitudinal modulus (tension)	$E_L{}^t$	2.07 Msi	14.3 GPa
Transverse modulus (tension)	$E_T{}^t$	1.82 Msi	12.5 GPa
Poisson's ratio	ν_{LT}	0.218	0.218
Shear modulus[a]	G_{LT}	0.86 Msi	5.9 GPa
Longitudinal modulus (compression)	$E_L{}^c$	1.91 Msi	13.2 GPa
Transverse modulus (compression)	$E_T{}^c$	1.75 Msi	12.1 GPa
Longitudinal strength (tension)	$S_L{}^t$	27.6 ksi	190.3 MPa
Transverse strength (tension)	$S_T{}^t$	25.4 ksi	175.1 MPa
Longitudinal strength (compression)	$S_L{}^c$	35.8 ksi	246.8 MPa
Transverse strength (compression)	$S_T{}^c$	34.1 ksi	235.1 MPa
Shear strength[a]	S_{LT}	9.0 ksi	62.0 MPa

[a]From Ref *17*.

dent of bolt diameter, increasing torque resulted in a corresponding increase in clamping pressure, but the magnitude was lower as fastener diameter (D) increased. Larger diameters result in lower clamping pressures because of the changes in fastener cross sectional area and the area of the constraining washer. All three relationships were approximately linear in the 0 to 52.0 MPa (7500 psi) clamping pressure range, but data for the 6.35-mm bolt extend beyond this range and become nonlinear. This torque versus clamping pressure data was used to determine applied torques necessary to keep clamping pressure constant while varying fastener diameter.

The bolted joint experiments independently investigated the effects of clamping pressure, fastener diameter, e/D, and W/D on the strength and failure mode behavior. Results from these parameter studies revealed important information about behavior peculiar to SMC bolted joints.

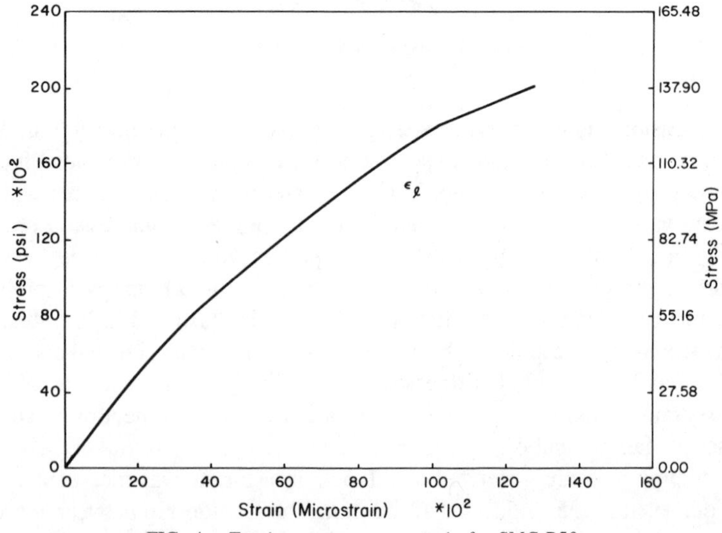

FIG. 4—*Tension stress versus strain for SMC-R50.*

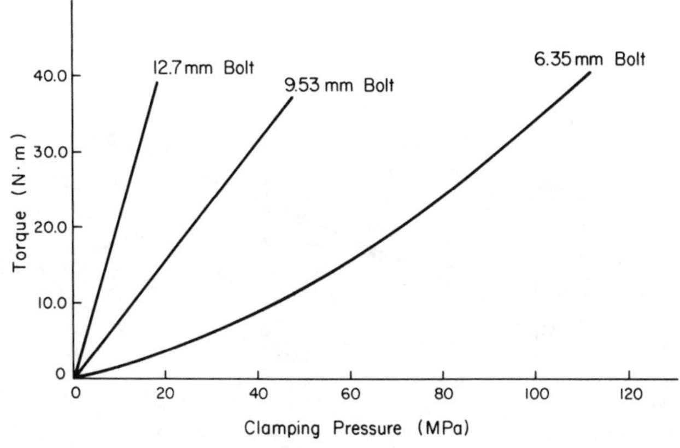

FIG. 5—*Plot of torque versus clamping pressure for 6.35, 9.53, and 12.70-mm bolts.*

Load-deflection data was recorded for each test, and the characteristics of the two typical responses are important. Although schematic, the two curves shown in Fig. 6 illustrate the salient features of the load deflection behavior of the SMC-R50 double lap bolted joints. Curve A shows the response for a specimen which exhibited net tension failure. Notice that nonlinear behavior developed shortly before failure. Curve B illustrates the response for a specimen which exhibited a combination cleavage-net tension failure. The cleavage failure initiation caused a small load drop after which reloading occurred until final catastrophic failure. Again there is evidence of nonlinear behavior near the failure loads. This nonlinear behavior is significant in terms of understanding the failure mechanism.

Since clamping pressure has been shown to be important in continuous fiber composite bolted joints [*18*], the effect of clamping pressure on strength and failure mode was assessed for the SMC system. Clamping pressures ranged from 0 (pin bearing) to approximately 79.2 MPa (11 500 psi) for the 6.35-mm bolts

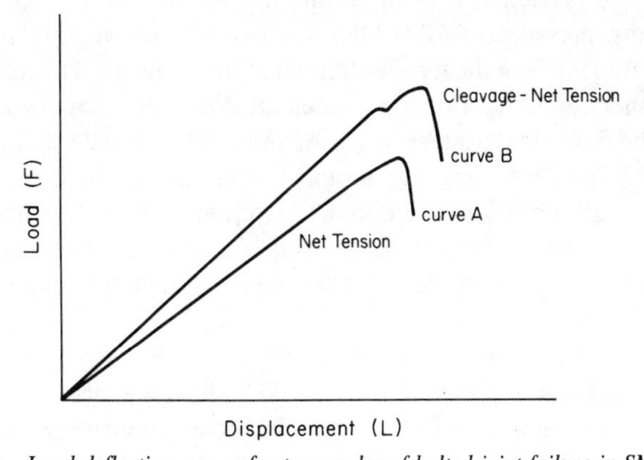

FIG. 6—*Load deflection curves for two modes of bolted joint failure in SMC-R50.*

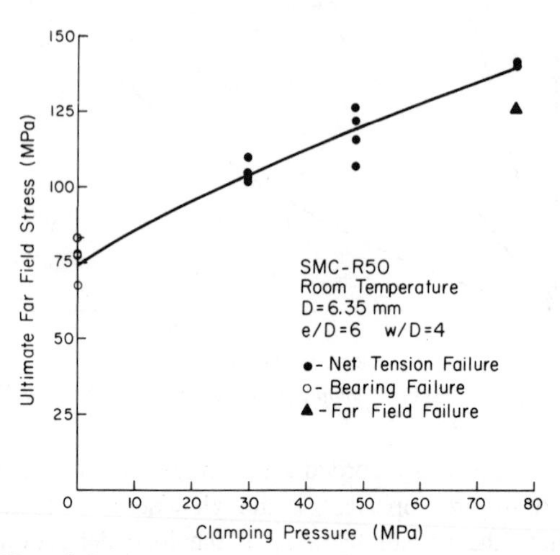

FIG. 7—*Effect of clamping pressure on bolted joint strength, 6.35-mm fastener.*

and between 0 to 31.0 MPa (4500 psi) for the 9.53-mm bolt. The results are shown in Figs. 7 and 8. The pin bearing tests resulted in bearing failures for both fastener sizes; all other failures were net tension. For the 6.35-mm fastener, strength increased linearly with increasing clamping pressure. The variance of the results was low, and one far-field failure was observed. A far-field failure is a tension failure which occurred in the specimen test section at a location other than through the joint, a characteristic of SMC composites. Results for the 9.53-mm fastener exhibited much larger scatter. The increase in strength with clamping pressure appeared to level off as the 30-MPa range was reached. Other than the occurrence of a far-field failure, the trends in these results are similar to those for continuous fiber systems. It is interesting to note that for the 6.35-mm fastener at a clamping pressure of 70.0 MPa (11 000 psi) the strength of the joint is approximately 79.0% of the tensile strength of the laminate. The reduction in area for the fastener cutout, ignoring any potential stress concentration effects, would result in a far-field tensile stress of 75.0% of the SMC tensile strength. Thus, the loss in strength of mechanically fastened joints in the SMC material for this geometry at high torques is due only to reduction in the net section area. There appears to be no stress concentration penalty for the 6.35-mm fastener; in fact the friction due to clamping appears to make the joint slightly stronger than allowed by the material alone.

A hole size effect has been documented for notched strength of SMC composites [*19*] and was expected to exist for SMC bolted joints. Results confirmed the sensitivity of strength on fastener size, but other geometric parameters appear to influence the relationship between strength, fastener size, and failure mode. In

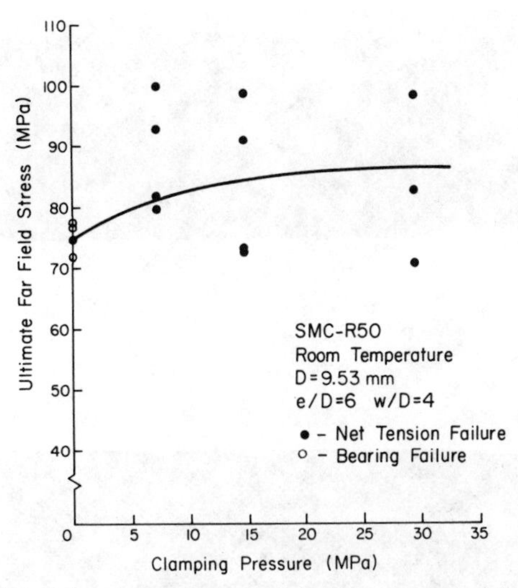

FIG. 8—*Effect of clamping pressure on bolted joint strength, 9.53-mm fastener.*

Fig. 9 results for the geometry $e/D = 2$ and $W/D = 6$ are presented. Strength was observed to decrease with increase in fastener size. All but one sample failed in the combination cleavage-net tension mode. Cleavage failure initiated first as a tension failure on the specimen edge and propagated toward the fastener until the net tension mode combined to produce catastrophic failure (Fig. 10). For geometries of $e/D = 4$ with W/D ranging between 2 and 4 a similar trend of decreasing strength with increasing fastener diameter was observed. Larger W/D

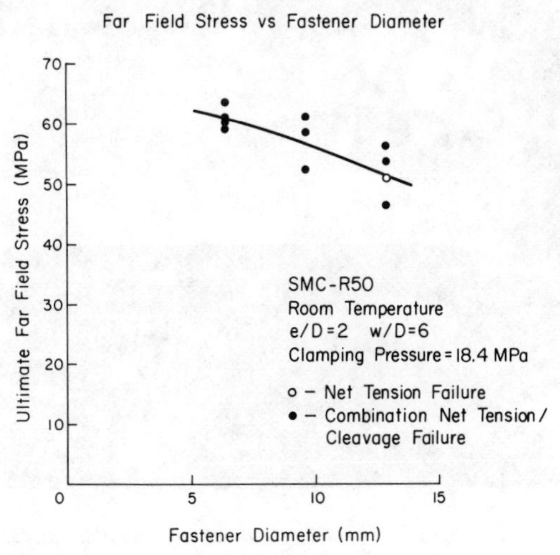

FIG. 9—*Influence of fastener size on bolted joint strength in SMC-R50,* e/D = 2, W/D = *6.*

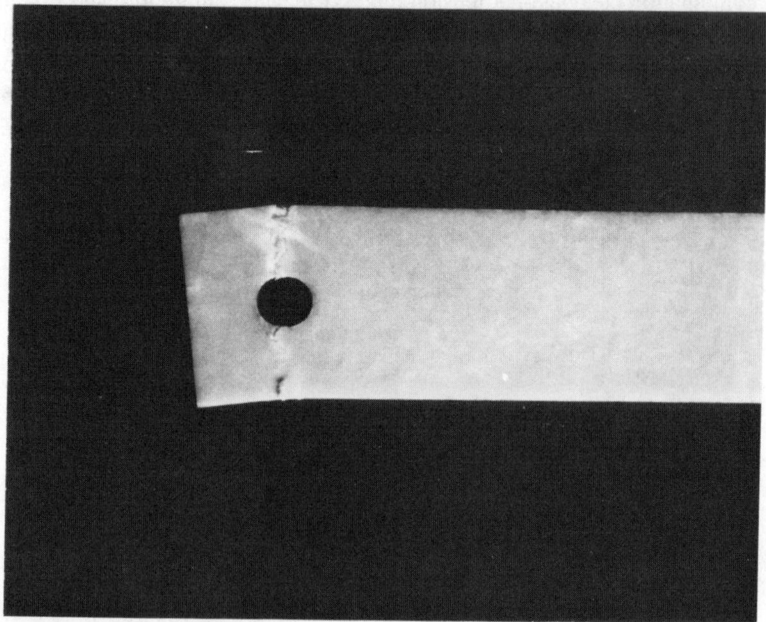

FIG. 10—*Photograph of cleavage and net tension failure modes.*

appeared to result in a more pronounced decrease with increasing D, although scatter was large enough to prevent certainty in establishing this trend. The results for $e/D = 6$, $W/D = 4$ revealed an apparent reversal (see Fig. 11) of the relationship between fastener size and strength. For this geometry, strength appeared to increase with increasing fastener size. Again, scatter in the data is large enough to obscure the details of the trend. Net tension failures predominated except for the $e/D = 2$ geometry, so a change in failure mode was not responsible for the apparent reversal in the strength versus fastener size trend. Dependent upon the geometry, the magnitude of the decrease in strength over the range of fastener sizes investigated was between 3 and 15%.

The effects of W/D and e/D on strength are summarized by Figs. 12 and 13, respectively. Increases in W/D resulted in increased strength. This result is in opposition to results normally encountered in continuous fiber composite systems where the bearing failure mode occurs. The main reason for the SMC strength trend was that the failure mode did not shift from net tension to bearing with increase in W/D. SMC is softer and less anisotropic resulting in lower stress concentrations translating to strength retention. Figure 12 shows results for a 6.35-mm fastener; similar results were found for the other two diameters.

Strength increased with increasing e/D in the range between 2.0 and 6.0 and appeared to remain constant for values greater than 6.0. This is similar to trends observed for continuous fiber systems. Failure mode shifted from combination cleavage-net tension to net tension for e/D greater than 2.0. No bearing failures were recorded at large e/D, even when W/D was 6.0, as would be expected for continuous fiber systems.

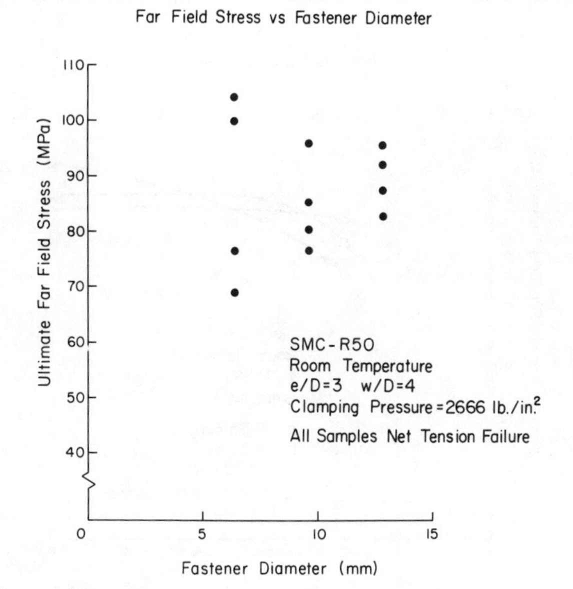

FIG. 11—*Influence of fastener size on bolted joint strength in SMC-R50,* e/D = *6,* W/D = *4.*

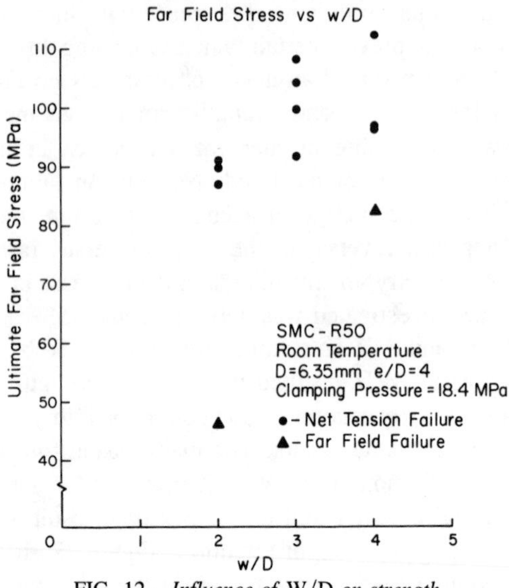

FIG. 12—*Influence of* W/D *on strength.*

Conclusions

This investigation of the strength and failure mode behavior of SMC-R50 has revealed several important characteristics of bolted joint performance which derive from the nature of the material. The stress-strain behavior is nonlinear, and this nonlinear behavior was observed in load versus deflection behavior for bolted joints. In general, scatter in the data was high enough that four data points were

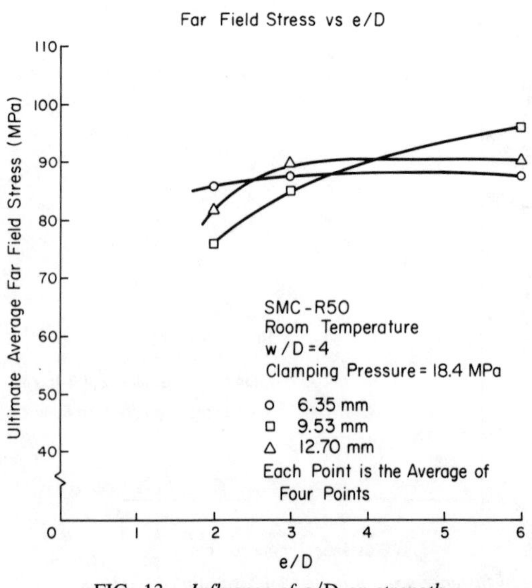

FIG. 13—*Influence of* e/D *on strength.*

insufficient to make conclusive inferences from the data. This large scatter was likely attributable to the variability of microstructure in relation to fastener hole placement, not to the test method or material variability alone. Several failures were recorded in the far-field test section, away from the fastener penetration. This phenomenon, specific to SMC type materials, has been documented in open-hole, notched strength tests [*19*]. Until the penetrating hole reaches a critical size, the "inherent flaw" characteristics of the SMC provide a probability of failure in the unperturbed test section. Evidence indicated that for a 6.35-mm fastener, high clamping in combination with the approximate in-plane isotropy resulted in an apparent neutralization of stress concentration effects on the joint strength. It is therefore possible to design a joint in SMC which has a strength loss due only to the reduction in load carrying area of the material. Unfortunately, regardless of geometry, catastrophic tension failure modes prevailed, an undesirable characteristic for strength critical designs.

References

[*1*] Hart-Smith, L. J., "Bolted Joints in Graphite-Epoxy Composites," NASA Contract Report NASA CR–144899, Douglas Aircraft Co., Long Beach, CA, June 1976.

[*2*] Van Siclen, R. C. in *Proceedings,* Army Symposium on Solid Mechanics, AMMRC MS74–8, Army Mechanics and Materials Research Center, Watertown, MA, Sept. 1974, pp. 120–138.

[*3*] Collins, T. A., *Composites,* Vol. 8, Jan. 1977, pp. 43–55.

[*4*] Hyer, M. W. and Lightfoot, M. C. in *Composite Materials: Testing and Design (Fifth Conference), ASTM STP 674,* S. W. Tsai, Ed., American Society for Testing and Materials, Philadelphia, 1979, pp. 118–136.

[*5*] Quinn, W. J. and Matthews, F. L., *Journal of Composite Materials,* Vol. 11, April 1977, p. 139.

[*6*] Wilson, D. W. and Pipes, R. B., "Behavior of Composite Bolted Joints at Elevated Temperature," NASA Contract Report 159137, University of Delaware Center for Composite Materials, Newark, Sept. 1979.

[*7*] Wilson, D. W., Pipes, R. B., Webster, J. W., and Riegner, D. L. in *Test Methods and Design Allowables for Fibrous Composites, ASTM STP 734,* American Society for Testing and Materials, Philadelphia, 1981, pp. 195–207.

[*8*] Kim, R. Y. and Whitney, J. M., *Journal of Composite Materials,* Vol. 10, April 1976, p. 149.

[*9*] Wilkins, D. J. in *Composite Materials: Testing and Design (Fourth Conference), ASTM STP 617,* American Society for Testing and Materials, Philadelphia, 1977, pp. 497–513.

[*10*] Ramkumar, R. L. in *Test Methods and Design Allowables for Fibrous Composites, ASTM STP 734,* American Society for Testing and Materials, Philadelphia, 1981, pp. 376–395.

[*11*] Waszczak, J. P. and Cruse, T. A., *Journal of Composite Materials,* Vol. 5, July 1971, p. 421.

[*12*] Waszczak, J. P. and Cruse, T. A., "A Synthesis Procedure for Mechanically Fastened Joints in Advanced Composite Materials," AIAA/ASME/SAE 14th Structures, Structural Dynamics, and Materials Conference, Williamsburg, VA, March 1973.

[*13*] Soni, R. Som, "Failure Analysis of Composite Laminates with a Fastener Hole," Technical Report AFWAL-TR-80-4010, Materials Laboratory, Air Force Wright Aeronautical Laboratories, Dayton, OH, March 1980.

[*14*] Eisenmann, J. R., "Bolted Joint Static Strength Model for Composite Materials," NASA-TM-X-3377, National Aeronautics and Space Administration, Washington, DC, 1976.

[*15*] Wilson, D. W. and Pipes, R. B. in *Proceedings,* International Conference on Composite Structure, I. H. Marshall, Ed., Applied Science, London, 1981, pp. 34–49.

[*16*] York, J. M., Wilson, D. W., and Pipes, R. B., *Journal of Reinforced Plastics and Composites,* Vol. 1, April 1982, p. 141.

[*17*] Denton, D., "The Mechanical Properties of an SMC-R50 Composite," Owens-Corning Fiberglas Corp., Granville, OH, 1979.

[*18*] Stockdale, J. H. and Matthews, F. L., *Composites,* Vol. 7, Jan. 1976, pp. 34–38.

[*19*] Yost, B. A., "Predicting the Notched Strength of Composite Materials," M.MAE thesis, Dept. of Mechanical and Aerospace Engineering, University of Delaware, Newark, Aug. 1981.

John W. Gillespie, Jr., [1] *and R. Byron Pipes* [2]

Thermoelastic Response of the Cylindrically Orthotropic Disk — Numerical and Experimental Evaluation

REFERENCE: Gillespie, J. W., Jr. and Pipes, R. B., **"Thermoelastic Response of the Cylindrically Orthotropic Disk — Numerical and Experimental Evaluation,"** *High Modulus Fiber Composites in Ground Transportation and High Volume Applications, ASTM STP 873,* D. W. Wilson, Ed., American Society for Testing and Materials, Philadelphia, 1985, pp. 86–102.

ABSTRACT: Transfer molding of axisymmetric bodies with fiber-reinforced molding compounds through gates coincident with the geometric axis of revolution (z) results in fiber orientation distributions which may also be independent of tangential position (θ). In the present study, the thermoelastic response of the annular disk is correlated with analytic, numerical integration and finite-element predictions. Methodology based upon microscopic examination of the disk fiber orientation distribution is presented for the material property characterization required by the various solution techniques. Correlation with experimental results indicates that all three techniques adequately predict the in-plane deformation and strain response of the cylindrically orthotropic disk. The approximate elasticity solution of the layered fiber orientation distribution obtained by the finite-element method reveals an interlaminar stress state at tractionfree surfaces which decays within one disk thickness. Consequently, the analytical and numerical integration techniques which assume homogeneity through the disk thickness are not applicable for stress calculations in the vicinity of the free edges.

KEY WORDS: transfer molding, short fiber thermosets, process-induced fiber orientation, cylindrically orthotropic, thermoelastic material properties, interlaminar stress

Transfer molding of axisymmetric bodies with fiber-reinforced molding compounds through gates coincident with the geometric axis of revolution (z) results in fiber orientation distributions which may also be independent of tangential position (θ). The fiber orientation distribution is determined by mold geometry, process conditions, and the rheological properties of the molding material, and in general will vary throughout an r-z plane passing through the structure. The influence of microstructure on the thermoelastic properties of composite materi-

[1]Associate scientist, and [2]professor and director, respectively, Center for Composite Materials, Department of Mechanical and Aerospace Engineering, University of Delaware, Newark, DE 19711.

als has been investigated by McCullough, Jarzebski, McGee, and Mukhopadhyay [*1*,*2*,*3*], who have developed micromechanics models [*4*] for the prediction of thermoelastic properties of short fiber-reinforced composites as a function of composition, degree of orientation of the reinforcement, and the effective aspect ratio of the reinforcement. Consequently, a center gated axisymmetric body will be cylindrically orthotropic with the pointwise variation of elastic properties determined uniquely by the fiber orientation distribution.

In the present study, the thermoelastic response of the cylindrically orthotropic disk is investigated. The transfer mold configuration employed in the processing study is presented in Fig. 1. The material system consisted of a short glass fiber-reinforced phenolic thermoset of two viscosities, both having a fiber volume fraction of 40%. Processing studies (in conjunction with optical microscopy) by Taggart [*5*] and Gillespie [*6*] and numerical predictions by Givler [*7*] and York [*8*] reveal basic orientation effects that can be applied to simple geometries, as illustrated in Fig. 2. In the disk mold geometry, the dominant mechanism is diverging flow, which orients fibers in the circumferential direction. Secondary mechanisms [*6*] which align fibers in the radial direction are the shear stress distribution through the thickness of the mold cavity and the squeeze flow phenomenon which occurs in the final stages of the mold fill process. Optical microscopy of the disk cross section is employed to characterize the fiber orientation distribution directly for both material systems. The typical fiber orientation distribution is shown schematically in Fig. 3, where the boundary layer thickness of radially oriented short fibers attains a maximum before diminishing at the outer diameter. The microstructure resembles a 3-ply laminate, [0/90/0], of radial (0° layer) and circumferential (90°) orientations. This "laminate analogy" is instrumental in the evaluation of the thermoelastic response of the cylindrically orthotropic disk.

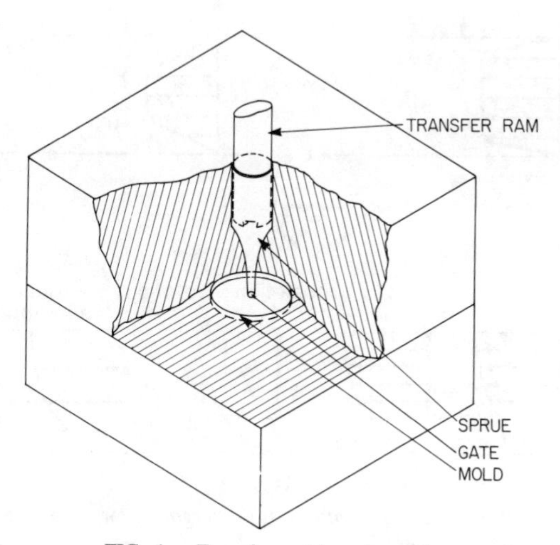

FIG. 1—*Transfer mold configuration.*

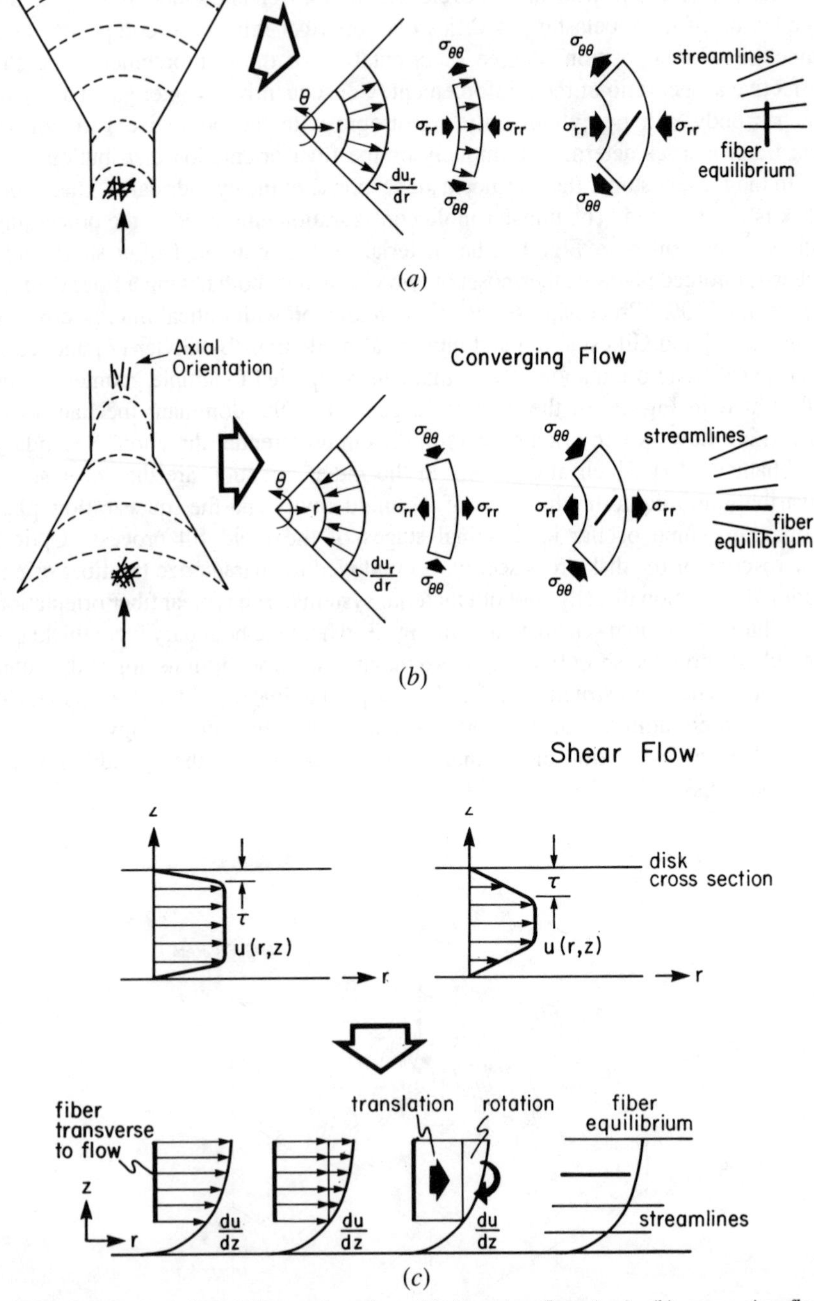

FIG. 2—*Basic rules of fiber orientation:* (a) *diverging flow* (top), (b) *converging flow* (center), *and* (c) *shear flow* (bottom).

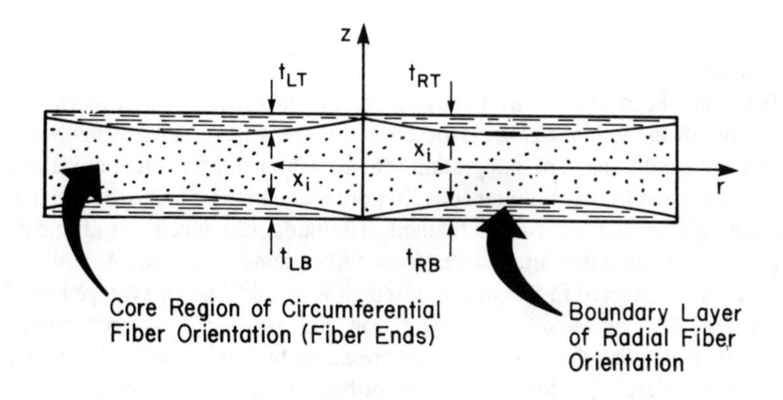

Total Boundary Layer Thickness

Optical Microscopy: Pointwise Characterization

$$t_e(x_i) = \frac{[t_{RT}(x_i) + t_{RB}(x_i)] + [t_{LT}(x_i) + t_{LB}(x_i)]}{2}$$

Non-Linear Least Squares Algorithm: Continuous Representation

$$t_e(r) = \sum_{n=0}^{4} a_N r^N$$

Average Total Boundary Layer Thickness

$$t_e^{AV} = \frac{\int_a^b t_e(r)dr}{b-a}$$

FIG. 3—*Typical fiber orientation distribution and boundary layer characterization.*

In general, the cylindrically orthotropic disk, homogeneous through the thickness, will still develop thermal residual stresses due to the mismatch in the thermal coefficients of expansion in the radial and circumferential directions upon cooling to ambient conditions from process temperatures [*9*]. In the present study, additional residual stresses result from the interaction of the 0 and 90° layers which may initiate transverse cracking in the boundary layer of radial fiber orientation [*10*]. Since boundary layer thicknesses can be significantly influenced by process conditions, analysis techniques are required which will enable the molder to circumvent potential problems with radial cracking in molded parts. Analytic solutions do not exist for arbitrary material property distributions. Consequently, finite-element techniques are required to provide approximate elasticity solutions. Employing lamination theory to calculate effective properties simplifies the problem formulation significantly. However, the cylindrically orthotropic material property distribution, although homogeneous, will remain a function of radial position in the disk geometry. The governing equation in terms of displacements yields a linear second order ordinary differential equation with nonconstant coefficients. Closed form solutions do not exist for material property distributions characteristic of transfer molded disks. A numerical integration scheme has been utilized by Gillespie [9] for the analysis of cylindrically orthotropic disks with variable elastic constants for a large class of loadings and boundary conditions. The integration scheme requires the reduction of the governing equation to two simultaneous first order equations which are solved using Hamming's predictor-corrector method. The numerical scheme is efficient and represents an attractive alternative to the finite-element method. A further simplification considered in the present research is to calculate an average boundary layer thickness of radial fiber orientation and associated material properties (Fig. 3). Analytic solutions have been presented by Lekhnitskii [11], Nowinski [12], and Gillespie [9] for cylindrically orthotropic disks with constant homogeneous properties. The three analysis techniques discussed represent varying degrees of sophistication, complexity, and applicability. Obviously, the simplest analysis which accurately models experimental observations is desirable in the design of transfer molded components. The applicability of the analyses presented will be determined by direct correlation with experimental data.

Results

The experimental effort consisted of the quantification of the thermoelastic response of the cylindrically orthotropic disk. Specimen configurations are presented in Fig. 4, where three annular disk geometries are machined from pristine disks. The molding material consisted of a short glass fiber-reinforced phenolic thermoset having a glass content of 40% by volume. Previous processing studies [5,6] indicate that fill rate and viscosity strongly influence the magnitude of the characteristic fiber orientation distribution (Fig. 3) by altering the velocity and shear stress profiles through the disk thickness, as illustrated in Fig. 2. Consequently, two viscosities, μ, designated Material 1 (μ_1) and Material 2 (μ_2) are considered in the present study ($\mu_2 > \mu_1$) to substantiate analysis techniques for process-induced variations in the fiber orientation distribution. Process conditions are summarized in Table 1.

In Fig. 4, the annular disk specimen geometries and strain gage locations are presented. A minimum of three tests were performed to evaluate the thermoelastic response of the cylindrically orthotropic disk. Specimens were subjected to a linear heating profile to a maximum temperature of 130°C (175°F). Radial and tangential strain component measurements were taken at regular increments in the surface temperature. Surface temperature is measured directly with a temperature strain gage. Linear response is observed over the temperature range investigated for all specimens. Experimental results are summarized for Materials 1 and 2 in Table 2.

The fiber orientation distribution is determined directly by diametrically sectioning and polishing specimens after measuring the thermoelastic response. Optical microscopy is then employed to determine the pointwise variation in thickness of radial fiber orientation. In Figs. 5 and 6, the distribution of total

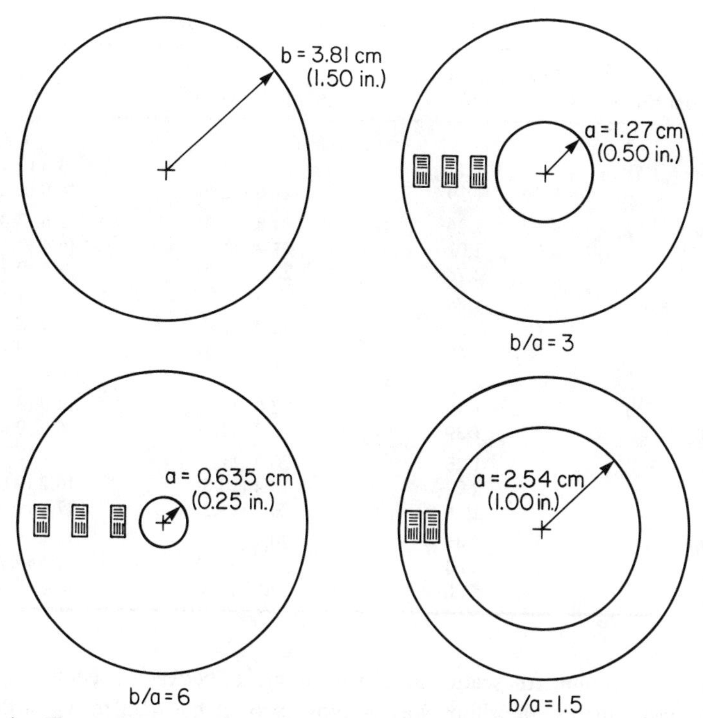

FIG. 4—*Thermoelastic response of annular disks: specimen geometry. Disk thickness = 0.318 cm (0.125 in.).*

boundary layer thickness is presented for Materials 1 and 2, respectively. The total boundary layer thickness, t_e, is defined in Fig. 3 and is consistent with the assumption of axisymmetry. Test specimens molded under identical process conditions yield fairly reproducible boundary layer profiles for a short fiber

TABLE 1—*Process conditions.*[a]

Specimen No.	Time to Fill, s	Pressure, MPa/psi	Mold Temperature, °C (°F)	Preform Temperature, °C (°F)
		MATERIAL 1 (70% ORIFICE FLOW)		
8	20	4.1 (600)	168 (335)	116 (240)
9	10	4.1 (600)	168 (335)	116 (240)
10	21	4.1 (600)	168 (335)	116 (240)
11	19	4.1 (600)	168 (335)	116 (240)
		MATERIAL 2 (60% ORIFICE FLOW)		
61	18	8.3 (1200)	168 (335)	116 (240)
62	19	8.3 (1200)	168 (335)	116 (240)
64	12	8.3 (1200)	168 (335)	116 (240)
65	16	8.3 (1200)	168 (335)	116 (240)

[a]Gate diameter = 0.318 cm (0.125 in.).

TABLE 2—*Experimental results: thermoelastic strain response.*

Specimen No.	Strain Gage Location, r/a	$\varepsilon_r/\Delta T$, $\mu\varepsilon$/°C ($\mu\varepsilon$/°F)	$\varepsilon_\theta/\Delta T$, $\mu\varepsilon$/°C ($\mu\varepsilon$/°F)
	MATERIAL 1		
64 (b/a = 1.5)	1.13	25.9 (14.4)	13.3 (7.4)
	1.39	21.6 (12.0)	14.8 (8.2)
62 (b/a = 3.0)	1.26	24.5 (13.6)	12.6 (7.0)
	2.02	21.4 (11.9)	16.4 (9.1)
	2.78	23.0 (12.8)	17.5 (9.7)
9 (b/a = 6.0)	1.45	25.4 (14.1)	7.9 (4.4)
	3.64	22.9 (12.7)	15.7 (8.7)
	5.60	22.5 (12.5)	16.9 (9.4)
	MATERIAL 2		
64 (b/a = 1.5)	1.13	22.0 (12.2)	14.9 (8.3)
	1.39	20.2 (11.2)	16.9 (9.4)
62 (b/a = 3.0)	1.26	20.3 (11.3)	... ...
	2.02	19.4 (10.8)	16.2 (9.0)
	2.78	20.2 (11.2)	17.8 (9.9)
61 (b/a = 6.0)	1.45	20.2 (11.2)	14.8 (8.2)
	3.64	20.2 (11.2)	17.5 (9.7)
	5.60	23.9 (13.3)	17.8 (9.9)

system. To account for scatter in t_e measurements between specimens, a nonlinear least squares algorithm was employed to fit the data to a fourth order polynomial. Profiles are superimposed on the experimental measurements.

Before correlation of experimental data with analytic and numerical predictions can be performed, thermoelastic properties of the circumferential and radial layers found in the disk microstructures must be evaluated. In Fig. 7, tension specimens are machined from Specimens 8 and 65 for Materials 1 and 2, respectively. Two independent measurements of modulus (E_r) and Poisson's ratio ($\nu_{r\theta}$) are determined at two radial positions indicated schematically in Fig. 7. Experimental results are summarized in Table 3, where values of E_r, $\nu_{r\theta}$, and t_e at identical radial locations are averaged. Lamination theory, in conjunction with effective property and boundary layer measurements at strain gage locations, enables layer properties (E_r, $\nu_{r\theta}$) to be evaluated in a straightforward manner. In Table 4, representative properties for the radial and circumferential orientations, which yield excellent correlation with mechanical characterization results in Table 3 for both materials, are presented. Finite-element analysis, however, requires the complete characterization of elastic constants for the cylindrically orthotropic material system. Consequently, experimental data (Table 4) are utilized to calibrate the micromechanics model [*4*] to provide accurate estimates for all material descriptors needed in the various analysis techniques discussed in the introduction. In Table 5, predictions based upon representative constitutive properties for the glass/phenolic material system are presented for direct correlation with experimental data (Table 4). Calibration consists of varying the degree of

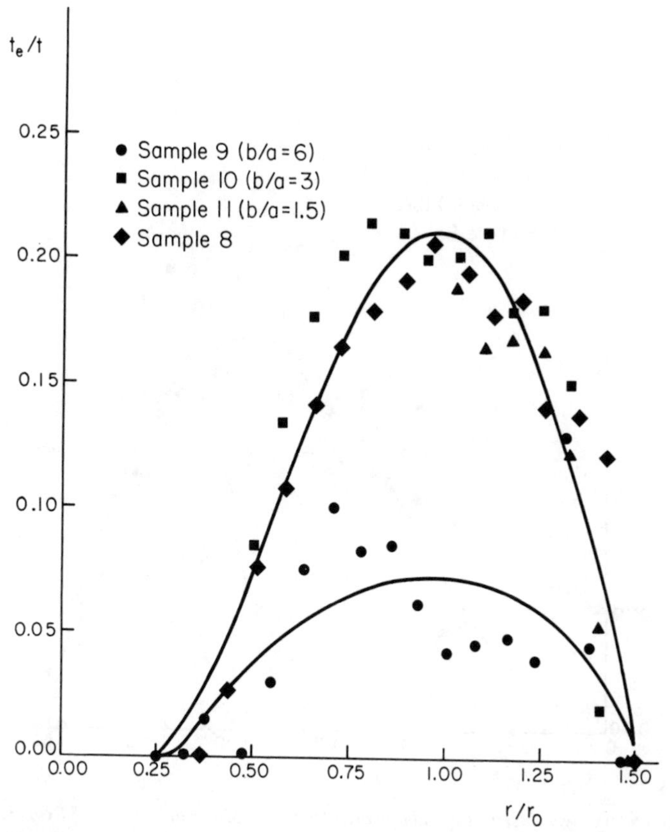

FIG. 5—*Boundary layer thickness of radially oriented fibers: Material 1* (r_0 = *2.54 cm [1.00 in.]).*

orientation to force agreement of the predicted longitudinal Young's modulus (E_L) with the experimental value in the radial direction (E_r). A value of Hermann's orientation parameter, f_p = 0.86 [*1,2,3*], provides excellent correlation with Poisson's ratio, $\nu_{r\theta}$, as well as with E_r, in both the radial (longitudinal Young's modulus) and circumferential (transverse Young's modulus) layers. Consequently, an equal degree of alignment of short fibers is predicted in the boundary layer of radial orientation and the core region of tangential fiber orientation. This agrees with microscopic examination of the disk microstructure and supports the laminate analogy which assumes the [0/90/0] configuration. Therefore, unidirectional lamina properties (radial orientation) are identified in Table 6 for the calculation of homogeneous properties for the analytic and numerical integration techniques.

A typical finite-element model of the annular disk specimen geometry is presented in Fig. 8. The variable boundary layer thickness of radial orientation defined by the polynomial fitted to experimental measurements is incorporated

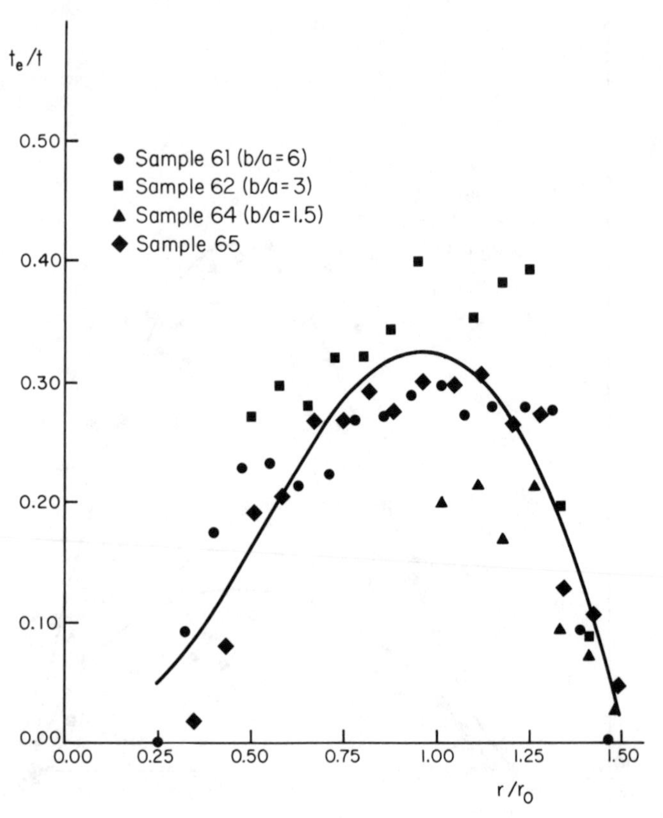

FIG. 6—*Boundary layer thickness of radially oriented fibers: Material 2* (r_0 = *2.54 cm [1.00 in.]*).

directly into the finite-element model. Material property input is based upon the micromechanics model predictions presented in Table 5 for the 0 and 90% orientations. Material property input for the analytic solution of the various annular disk geometries is summarized in Table 6 for Materials 1 and 2. An average boundary layer thickness is calculated and lamination theory is employed to evaluate homogeneous cylindrically orthotropic properties which are constant throughout the annular disk. The same methodology is pursued in the evaluation of material property distributions required by the numerical integration scheme. In Figs. 9 and 10, the typical variation of thermoelastic properties with radial position is presented. Inspection of the property distributions reveals the expected dependence on boundary layer thickness. Increasing t_e, for example, increases E_r as well as the thermal coefficient of expansion in the circumferential direction (α_θ). Conversely, E_θ and the radial coefficient of thermal expansion, α_r, are inversely proportional to the boundary layer thickness, t_e.

The layered fiber orientation distribution, in conjunction with the finite element solutions, reveals interlaminar stresses at the tractionfree surfaces. Analogous to the classicfree edge problem investigated by Pipes and Pagano [*13*], the

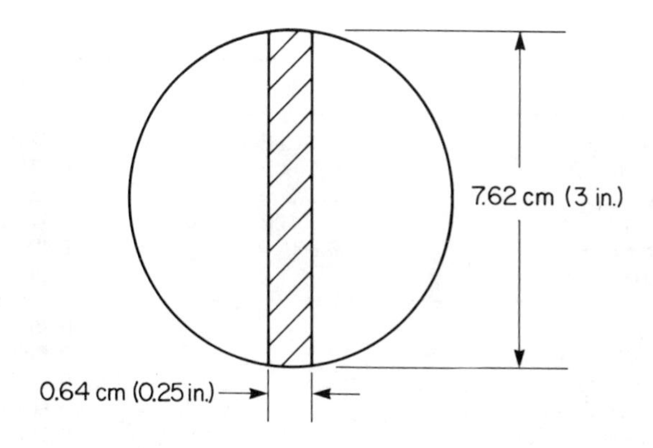

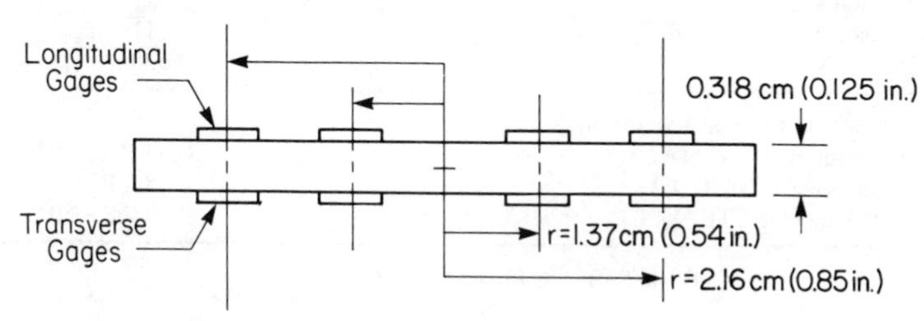

FIG. 7—*Mechanical characterization: tension specimen.*

TABLE 3—*Experimental results: mechanical characterization.*

	Strain Gage Location r, cm (in.)	Effective Modulus E_r, GPa (Msi)	Effective Poisson's Ratio, $\nu_{r\theta}$	Total Boundary Layer Thickness, Disk Thickness
		MATERIAL 1		
Specimen 8	1.37 (0.54)	12.13 (1.76)	0.21	0.09
	2.16 (0.85)	12.68 (1.84)	0.21	0.20
		MATERIAL 2		
Specimen 65	1.37 (0.54)	13.37 (1.94)	0.23	0.19
	2.16 (0.85)	13.71 (1.99)	0.24	0.31

TABLE 4—*Layer properties/experimental.*

	E_r, GPa (Msi)	$\nu_{r\theta}$
	MATERIAL 1	
Radial	19.98 (2.90)	0.310
Circumferential	11.02 (1.60)	0.171
	MATERIAL 2	
Radial	19.98 (2.90)	0.310
Circumferential	11.02 (1.60)	0.171

TABLE 5—*Thermoelastic properties*[a]*: finite element solution.*

Input Data	Resin	Fiber
Young's modulus (Pa)	0.4820E + 10	0.7235E + 11
Shear modulus (Pa)	0.1860E + 10	0.2715E + 11
Poisson's ratio	0.296	0.333
CTE (1/°C)	0.4050E − 04	0.5400E − 05
Volume fraction	0.600	0.400
Aspect ratio		50.00
Orientation		0.86
Longitudinal Young's modulus (Pa)		0.1988E + 11
Transverse Young's modulus (Pa)		0.1107E + 11
Perpendicular Young's modulus (Pa)		0.1107E + 11
2,3 shear modulus (Pa)		0.4343E + 10
1,3 shear modulus (Pa)		0.4674E + 10
1,2 shear modulus (Pa)		0.4674E + 10
1,2 Poisson's ratio		0.311
1,3 Poisson's ratio		0.311
2,3 Poisson's ratio		0.330
Coefficients of Thermal Expansion		
Longitudinal CTE (1/°C)		0.1331E − 04
Transverse CTE (1/°C)		0.2840E − 04
Perpendicular CTE (1/°C)		0.2840E − 04

[a]Center for Composite Materials Software Program: SMC.

TABLE 6A—*Thermoelastic properties: homogeneous solutions.*

Lamina	E_r, GPa	E_θ, GPa	$\nu_{r\theta}$	α_r, $\mu\varepsilon$/°C	α_θ, $\mu\varepsilon$/°C
Radial	19.98	11.02	0.310	13.5	27.9
Circumferential	11.02	19.98	0.171	27.9	13.5

TABLE 6B—*Radially averaged laminate properties.*

Specimen No.	E_r, GPa	E_θ, GPa	$\nu_{r\theta}$	α_r, $\mu\varepsilon$/°C	α_θ, $\mu\varepsilon$/°C	Average Total Boundary Layer Thickness, Disk Thickness
			MATERIAL 1			
9 (b/a = 6)	11.51	19.64	0.175	26.8	14.0	0.05
10 (b/a = 3)	12.47	18.67	0.184	24.5	14.9	0.16
11 (b/a = 1.5)	12.4	18.74	0.183	24.8	14.9	0.14
			MATERIAL 2			
61 (b/a = 6)	12.95	18.12	0.190	23.4	15.7	0.21
62 (b/a = 3)	13.16	17.98	0.192	23.0	16.0	0.24
64 (b/a = 1.5)	12.88	18.23	0.189	23.6	15.7	0.21

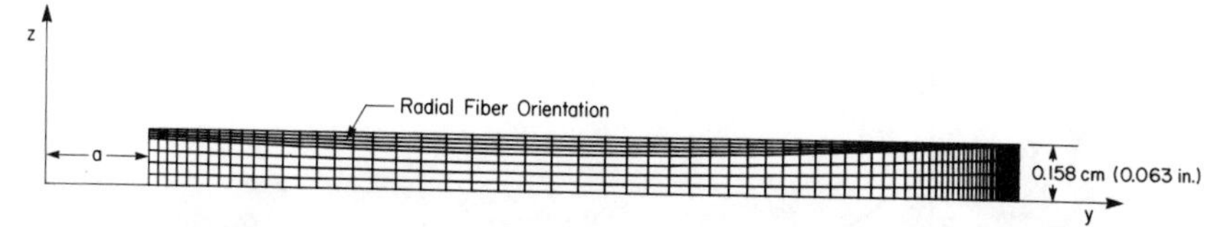

FIG. 8—*Typical finite element model of annular disk with boundary layer of variable thickness.*

interlaminar stress state for the cylindrically orthotropic disk also appears to diminish within one laminate (disk) thickness. Correlation of experimental results with the three analysis techniques is presented in Figs. 11–13. Excellent agreement is obtained for solutions of the tangential strain response for all annular disk geometries. The predicted strain response, ε_θ, is insensitive to the solution technique as well as the procedure for characterizing the material property distribution in the annular disk. The numerical integration and finite-element solutions diverge slightly from the analytic solution at the inner radius only. Consequently, the analytic solution may be applicable for deformation-critical designs, since the

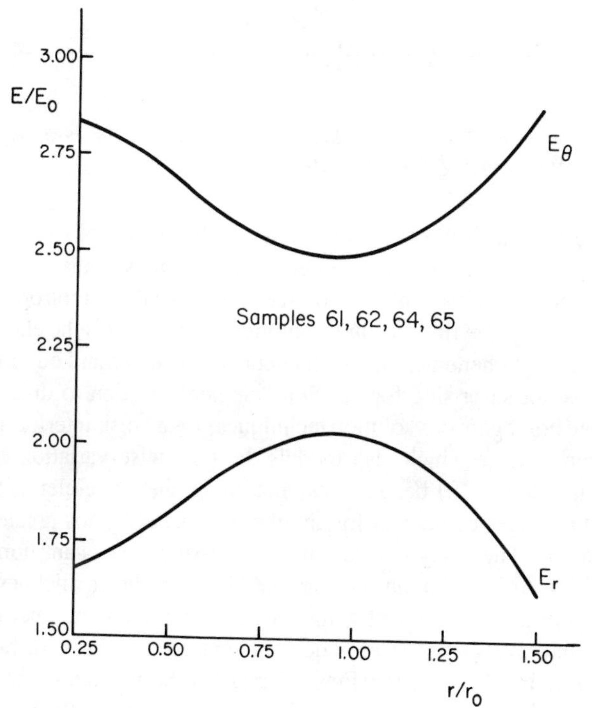

FIG. 9—*Variation of moduli with radial position: Material 2* (r_0 = *2.54 cm [1.00 in.]*, E_0 = *6.89 GPa [1.00 Msi]*).

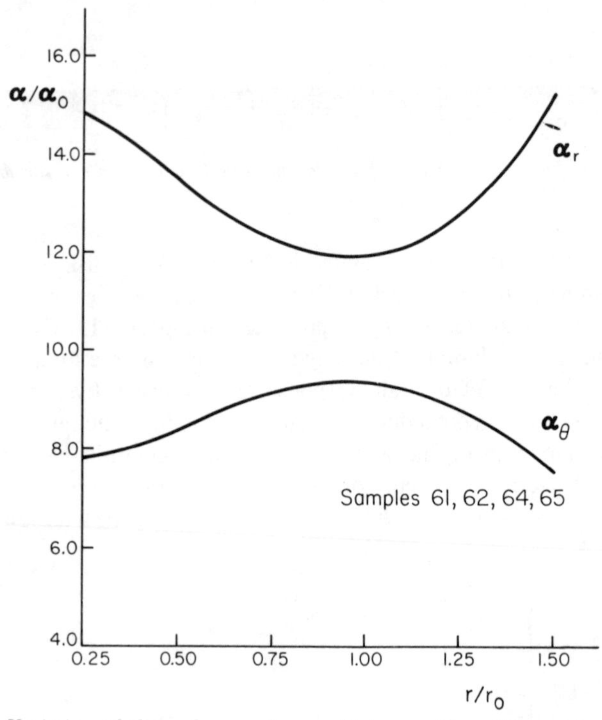

FIG. 10—*Variation of thermal expansion coefficients with radial position: Material 2 (r_0 = 2.54 cm [1.00 in.], $\alpha_0 = 1.8\ \mu\varepsilon/°C$ [1 $\mu\varepsilon/°F$]).*

radial displacement, u, is directly proportional to the tangential strain component ($u = r\varepsilon_\theta$) [8]. The radial strain component, ε_r, is highly sensitive to the solution technique employed in the analysis of the cylindrically orthotropic disk. As discussed previously, the finite-element solution approximates the elasticity solution of the free edge phenomenon which occurs at the inner and outer diameters. Therefore, it is not surprising for the finite-element solution to diverge significantly from the homogeneous solution techniques. In the disk interior, the numerical integration scheme which also models the pointwise variation of material properties (Figs. 9 and 10) becomes asymptotic to the finite-element solutions (Fig. 11) and becomes coincident for annular disk aspect ratios greater than 1.5 (Figs. 12 and 13). The analytic solution which assumes constant homogeneous properties (Table 6) based upon the average boundary layer thickness, t_e, also yields reasonably good agreement with finite-element and numerical integration results. All solution techniques provide excellent correlation with radial strain measurements on the disk interior (Figs. 12 and 13). Strain gages placed near the inner and outer diameters (Fig. 4) are influenced significantly by the severe gradients in ε_r which occur in the boundary layer of interlaminar stresses. Strain gages measure response averaged over the gage area, and consequently the largest discrepancy of experimental data with the three analysis techniques occurs

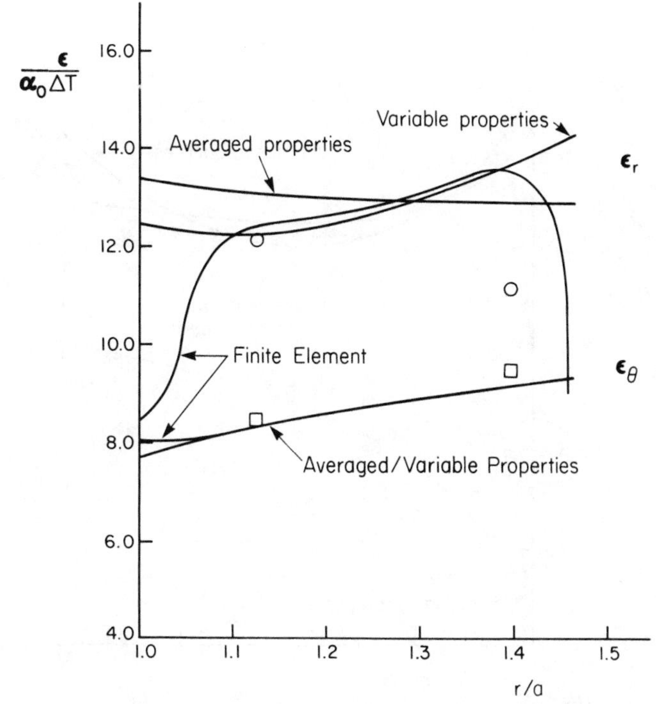

FIG. 11—*Correlation of results: Material 2, Specimen 64 (α_0 = 1.8 $\mu\varepsilon$/°C [1 $\mu\varepsilon$/°F]).*

in the free edge zones. Qualitatively, the finite-element method provides superior correlation with experimental data in the free edge regions.

In the context of the present formulation, the laminate analogy of the disk microstructure must be employed to calculate thermal residual stresses for the two solutions assuming homogeneity through the disk thickness. Therefore, strain components correspond to midplane strains, and lamination theory can be applied directly for the determination of layer stresses. The finite-element method yields thermal stresses in the radial and circumferential layers directly and verifies the above methodology for calculation of in-plane thermal stresses in regions not influenced by the free edge phenomenon. Consequently, the analytical and numerical solutions are not applicable in the free edge region, and one must revert to the complexities of the finite-element method to provide accurate estimates of free edge stresses.

Conclusions

In the present study, the thermoelastic response of a transfer molded axisymmetric disk through a center gate coincident with the geometric axis of revolution is experimentally determined for direct correlation with analytic and numerical predictions. Methodology is presented for the material property characterization

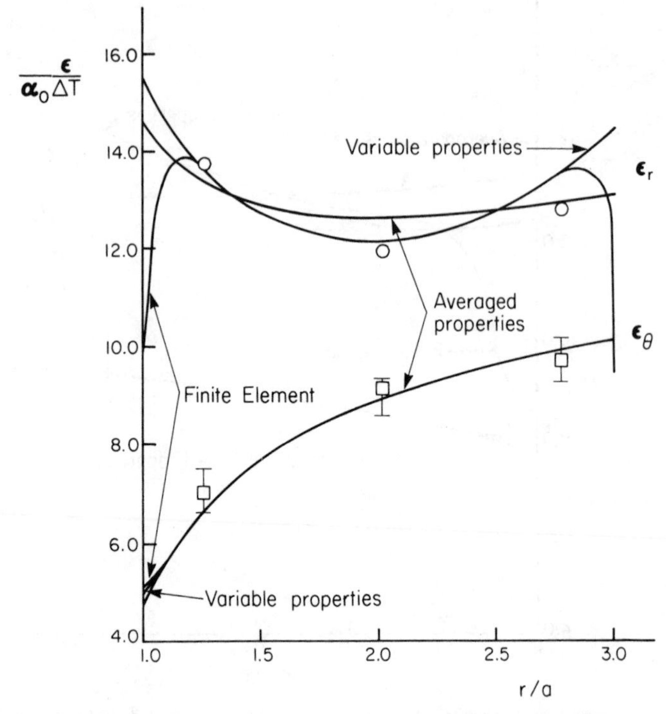

FIG. 12—*Correlation of results: Material 1, Specimen 10 (α_0 = 1.8 $\mu\varepsilon$/°C [1 $\mu\varepsilon$/°F]).*

for the analytic, numerical integration, and finite-element solutions based upon microscopic examination of the fiber orientation distribution. Correlation of experimental results indicate that all three techniques adequately predict in-plane deformation and strain response. The finite-element solution of the layered fiber orientation distributions reveals an interlaminar stress state at the tractionfree surfaces which decays within one disk thickness. Consequently, the analytic and numerical integration techniques which assume homogeneity through the disk thickness do not provide conservative stress calculations in the vicinity of the free edges.

Acknowledgments

This research was sponsored by the Rogers Corporation, Lurie Research and Development Center, Rogers, Connecticut. The authors gratefully acknowledge the assistance of Bruce Fitts and Vince Landi in defining program goals and in the molding of specimens.

References

[*1*] Mukhopadhyay, A. K., "On the Thermoelastic Response of Composite Materials," Report CCM-80-10, Center for Composite Materials, University of Delaware, Newark, 1980.

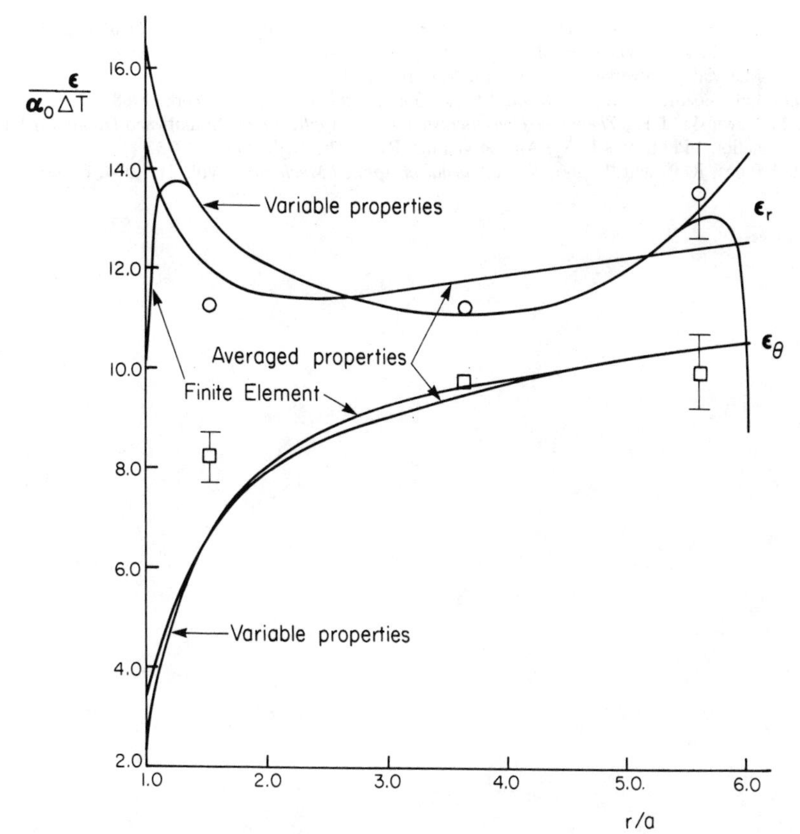

FIG. 13—*Correlation of results: Material 2, Specimen 61 (α_0 = 1.8 $\mu\varepsilon$/°C [1 $\mu\varepsilon$/°F]).*

[2] McGee, S. and McCullough, R. L., "Combining Rules for Predicting the Thermoelastic Properties of Particulate Filled Polymers, Polyblends and Foams," Report CCM-80-13, Center for Composite Materials, University of Delaware, Newark, 1980.

[3] McCullough, R. L., Jarzebski, G. J., and McGee, S. H., "Constitutive Relationships for Sheet Molding Materials," Report CCM-81-11, Center for Composite Materials, University of Delaware, Newark, 1981.

[4] SMC Computer Program: "Micromechanics Model for the Prediction of Thermoelastic Properties of Composite Materials," Center for Composite Materials Software, Center for Composite Materials, University of Delaware, Newark, 1983.

[5] Taggart, D. G. and Pipes, R. B., "Processing Induced Fiber Orientation in Transfer and Injection Molding," Report CCM-79-12, Center for Composite Materials, University of Delaware, Newark, 1979.

[6] Gillespie, J. W., Jr., Ellery, S. A., Mongan, D. T., and Givler, R. M., "Disk Processing Studies," Report in progress, Center for Composite Materials, University of Delaware, Newark, 1983.

[7] Givler, R. C., "Numerically Predicted Fiber Orientation in Dilute Suspensions," Report CCM-81-04, Center for Composite Materials, University of Delaware, Newark, 1981.

[8] York, J. L., "Fiber Orientation in Curvilinear Flow," Report CCM-82-05, Center for Composite Materials, University of Delaware, Newark, 1982.

[9] Gillespie, J. W., Jr., "Axisymmetric Analysis of the Cylindrically Orthotropic Disk of Variable Fiber Orientation," Report CCM-79-10, Center for Composite Materials, University of Delaware, Newark, 1979.

[10] Gillespie, J. W., Jr., "Thermal Response of the Cylindrically Orthotropic Disk with Through-the-Thickness Variation of Material Properties," Report CCM-79-20, Center for Composite Materials, University of Delaware, Newark, 1979.
[11] Lekhnitskii, S. G., *Anisotropic Plates,* Gordon and Breach, New York, 1968.
[12] Nowinski, J. L., *Theory of Thermoelasticity with Applications,* Sijthoff and Noordhoff International Publishers B. V., Alphen van den Rign, The Netherlands, 1978.
[13] Pipes, R. B. and Pagano, N. J., *Journal of Applied Mechanics,* Vol. 41, 1974, p. 668.

Ashok B. Hosangadi[1] *and Hong Thomas Hahn*[1]

Hygrothermal Degradation of Sheet Molding Compounds

REFERENCE: Hosangadi, A. B. and Hahn, H. T., **"Hygrothermal Degradation of Sheet Molding Compounds,"** *High Modulus Fiber Composites in Ground Transportation and High Volume Applications, ASTM STP 873,* D. W. Wilson, Ed., American Society for Testing and Materials, Philadelphia, 1985, pp. 103–118.

ABSTRACT: Environmental variables such as moisture and temperature degrade fiber-reinforced polymer composites to various levels. This paper characterizes the hygrothermal degradation of a sheet molding compound (SMC) containing 65% by weight of E-glass fibers. Six different environments were chosen for the study: room temperature (RT)/65% relative humidity (RH), RT/98% RH, RT/water, 75°C/65% RH, 75°C/65% RH, 75°C/98% RH, and 75°C/water. The extent of degradation was evaluated using visual, microscopic, and chemical analysis techniques. From this study, distinct color changes and blistering on the specimen surfaces were evident. Degradation was manifested in the ultrasonic attenuation and in the reduction of short beam strength and stiffness.

KEY WORDS: hygrothermal environments, degradation, swelling, sheet molding compounds, composite materials, moisture, strength, stiffness, ultrasonic attenuation, blisters

Fiber-reinforced polymer composites swell due to moisture absorption when exposed to humid environments. Elevated temperatures in moist environments induce hygrothermal strains resulting in dimensional changes in structures [*1*,*2*]. Such structural distortions produce internal stresses in addition to applied stresses.

Apart from introducing stresses and swelling, the moisture in polymer composite is also detrimental in other ways. Moisture sorption in the composite is mostly through the matrix via diffusion. It can also take place through cracks and voids in the composites and fiber-matrix interface [*1*]. Absorbed moisture plasticizes the matrix and lowers the glass transition temperature [*3*,*4*]. Entrapped water in voids and interfaces does not participate in the diffusion process but is "bound" to the system [*5*–*7*]. At elevated temperatures, bound moisture in voids and cracks expands, building osmotic pressure which can cause further cracking

[1]Graduate student and professor of mechanical engineering, respectively, Department of Mechanical Engineering, Washington University, St. Louis, MO 63130.

in the composite [*8–11*]. Over an extended period of time, the matrix material begins to degrade, resulting in surface discoloration, formation of blisters, and leaching out of low-molecular materials in the composites [*8*,*12*,*13*]. Such environmental degradation destroys polymer bonds and lowers the composite shear strength and stiffness [*14*,*16*].

While different types of fiber-reinforced composites exist, polyester-based randomly oriented short-glass fiber composites are receiving much attention and use [*17*,*18*]. The ease of manufacture of sheet molding compounds (SMC) partly relates to its popularity and potential for different uses [*19*,*20*]. Although the sorption behavior of SMC has been studied [*11*,*17*,*21–23*], its expansional properties and hygrothermal degradation are not fully known. Therefore, the objective of this paper is to characterize the hygrothermal degradation and swelling of SMC-R65 in various environments.

Experimental Procedure

Materials

The SMC composite chosen for this study is SMC-R65 (containing 65% by weight of E-glass fibers). The SMC panels were obtained from Owens-Corning Fiberglas Corp. The composition of the panels is listed in Table 1.

Moisture Absorption and Desorption

Table 2 lists the environments chosen for the present study. Saturated salt solutions of lead nitrate and magnesium nitrate were used to create relative humidities of 98 and 65%, respectively. Distilled and deionized water immersion was taken as 100% relative humidity (RH) environment. Saturated salt solutions were prepared using the same distilled and deionized water.

All specimens were 50 by 50-mm-square plates. Table 1 also shows the nominal thickness of specimens. Large planar dimensions of specimens were used to approximate one-dimensional through-the-thickness moisture sorption in the material.

Prior to environmental exposure, specimens were dried in a vacuum oven at 75°C. Specimens were considered fully dry when no apparent weight changes were noticed. The dimensions and dry weight of the specimens were recorded.

TABLE 1—*Composition of SMC-R65 (thickness of 4.5 mm).*

Function	Component	Manufacturer/Name	Weight %
Reinforcement	E-glass	OCF 433 AE-113	65
Resin	isophthalic polyester	OCF E987	31.8
Thickener	magnesium oxide	Hatco Modifier M	1.6
Catalyst	*tert*-butyl perbenzoate	Lucidol	0.32
Mold release	zinc stearate	Tennaco DLF-10	1.28

TABLE 2—*Environments studied.*

Temperature	Relative Humidity	Designation
Room temperature	65%	RT/65
	98%	RT/98
	Distilled water	RT/w[a]
75°C	65%	75/65
	98%	75/98
	Distilled water	75/w

[a]w = water.

After preconditioning, specimens were exposed to the environments listed in Table 2. Three specimens were used in each environment. During the moisture absorption cycle, specimen weight and dimensions were measured at regular time intervals. A Mettler balance with 0.001-g accuracy and a travelling microscope with 0.001-mm accuracy were used to monitor specimen weights and dimensions, respectively. Measurements were carried out until moisture equilibrium was reached.

The moisture desorption cycle was conducted at the same temperatures as in the absorption cycle in a vacuum oven until the specimens did not show any more weight change.

Short Beam Shear Tests

Rectangular beam specimens 8 mm wide by 30 mm long were prepared from virgin plates and moisture-conditioned plates, which had been fully saturated under various environments. After conditioning, one set of plates was kept in a laboratory environment for about three months before the test and the other set was completely dry.

The span-to-depth ratio used was 4. Based on simple beam theory, the maximum shearing stress is expected to be one eighth of the maximum bending stress. Testing was carried out at room temperature (RT) at a crosshead speed of 1.2 mm/min.

Material Damage Evaluation

Correlation between hygrothermal environment and the corresponding material degradation was evaluated based on (1) visual and microscopic observations of specimen surfaces and cross sections, (2) chemical analysis of leached out material, and (3) ultrasonic attenuation measurements using Sonoray 303B ultrasonic flaw detector with a 10 MHz transducer. While the first two techniques facilitated evaluation of material deterioration, the last technique helped to correlate mechanical property retention in material after environmental aging.

TABLE 3—*Diffusion parameters (SMC-R65).*

	$D \times 10^6$ mm²/s		$M\infty$, %	
SMC-R65	Absorption	Desorption	Absorption	Desorption
RT/65	0.88	1.19 (1.194)	1.40	1.36
RT/98	0.62	0.83 (0.829)	2.89	2.82
RT/w	0.60	1.05 (1.053)	2.96	2.78
75/65	5.12	6.03 (6.031)	1.33	1.36
75/98	3.20	2.82 (2.816)	3.12	2.90
75/w	2.74	2.91 (2.912)	3.35	3.03

Results and Discussion

From periodic specimen weight measurements, the diffusion coefficients and equilibrium moisture contents were calculated using the method described in Ref *21*, and the values are tabulated in Table 3. To compare the moisture diffusion behavior of a material in various environments, a nondimensional time parameter is defined as [*1*]

$$t^* = \sqrt{Dt}/h \tag{1}$$

where

D = diffusivity, mm²/s
t = time, s, and
h = specimen thickness, mm.

Figure 1 shows a master plot of moisture absorption in the material. At room temperature, the moisture diffusion is seen to follow the one-dimensional Fickian behavior described by [*21*]

$$\frac{M_t}{M_\infty} = 1 - \frac{8}{\pi^2}\sum_{j=0}^{\infty}\frac{1}{(2j+1)^2}\exp[-(2j+1)^2\pi^2 t^{*2}] \tag{2}$$

where M_t and M_∞ are the present and equilibrium moisture contents, respectively. At 75°C, however, the data deviate slightly from the Fickian behavior because of the damage to be discussed later.

If the temperature dependency of diffusivity is of Arrhenius type, then [*15,21,24*]

$$D = D_0 \exp(-E/\overline{R}\overline{T}) \tag{3}$$

where

D_0 = preexponential constant,

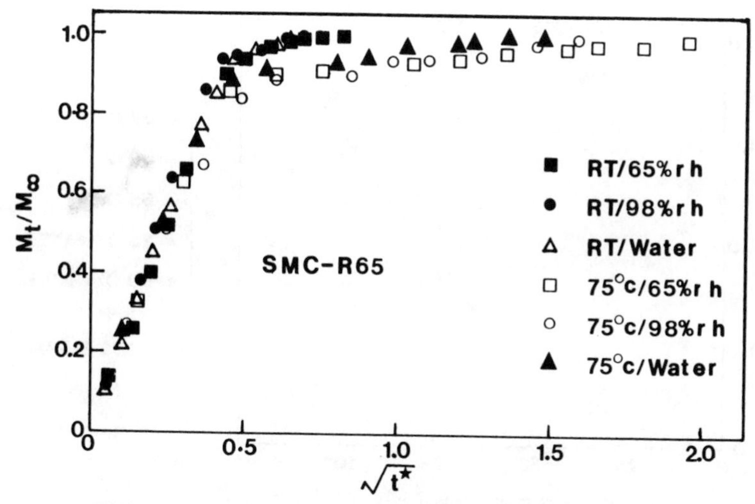

FIG. 1—*Moisture absorption in SMC-R65 in various environments.*

E = activation energy for moisture diffusion,
$\overline{R}$ = universal gas constant, and
$\overline{T}$ = absolute temperature.

An Arrhenius plot of the data in Fig. 2 indicates that the activation energy is fairly independent of relative humidity but the preexponential constant at the same temperature decreases with increasing relative humidity. The same was observed in Ref *21*.

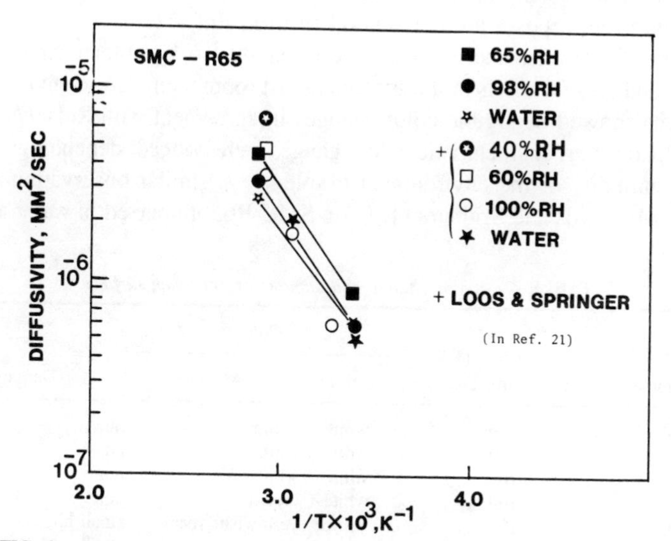

FIG. 2—*Diffusivity versus temperature during moisture absorption in SMC-R65.*

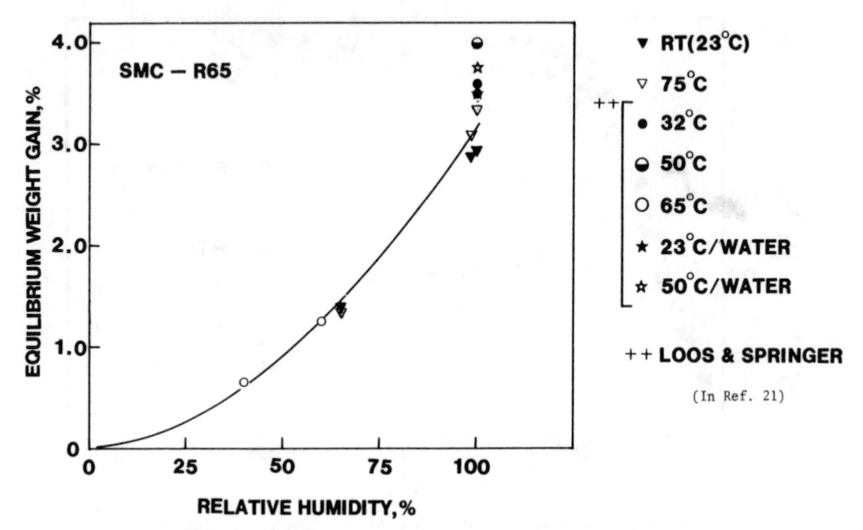

FIG. 3—*Equilibrium weight gain versus relative humidity.*

The equilibrium moisture content increases with relative humidity (RH) following a power law [*21*]

$$M_\infty = a\left(\frac{\mathrm{RH}}{100}\right)^b \tag{4}$$

The curve in Fig. 3 represents Eq 4 with $a = 0.057$ and $b = 1.700$. Since the experimental value of M_∞ only slightly depends on temperature, the data at both temperatures were combined and then fitted by Eq 4. Also shown in the figure for comparison purposes are the data taken from Ref *21*.

Effects of hygrothermal exposure were manifested by color changes on the surfaces and cross sections of the specimens. At room temperature environments, specimens showed negligible color changes in agreement with Refs *11* and *23*. At 75°C, however, characteristic color changes were noticed, depending upon the relative humidity of the environment (Table 4). A similar observation was reported by Edwards and Sridharan [*11*] on SMC-R65 immersed in water at 35°C.

TABLE 4—*Color change and damage on specimen surface.*

Environment	Period of Exposure, Days	Color		Damage
		Before	After	
RT/65	169	white	white	none
RT/98	169	white	white	none
RT/w	169	white	white	none
75/65	169	white	cream	none
75/98	169	white	yellowish cream	small blisters
75/w	169	white	whitish raw umber	small and large blisters

TABLE 5—*pH values of various solutions.*

Solution	pH Value[a]	
	Before Test	After Test
Deionized/distilled water	6.70	8.95 (RT)
Saturated $Pb(NO_3)_2$ salt solution in distilled water	3.70	3.17 (RT)
		5.45 (75°C)
Saturated $Mg(NO_3)_2$ salt solution in distilled water	6.20	5.40 (RT)
		6.30 (75°C)
Fluid extracted from blisters	. . .	3.70
		3.0[b]

[a]pH value of neutral water: 7.
[b]Reported value in Refs *8* and *30*.

While probing for the probable causes for such characteristic color changes on specimens, we noticed that at 75°C the salt solution of lead nitrate was decomposing, creating an acidic atmosphere in the environmental chambers [*25*]. A comparison with the water immersion specimens indicates that over an extended period of time the color of the specimens changed due to material degradation in acidic environment (Table 5).

Apart from color changes, certain localized microscopic features were noticed on the specimen surfaces. Blistering was noticed after nine to ten days of exposure to 75°C/98% RH and 75°C/water environments (Fig. 4). Similar observations were made by Edwards and Sridharan on SMC-R65 exposed to water immersion at 68°C [*11*]. Other investigators [*8–12*] reported blistering phenomenon in field and laboratory specimens and marine structures. Blisters ruptured upon finger pressure, oozing water and later leaving dark stains behind them [*11*,*23*].

Blisters in specimens appear due to concentration and expansion of water inside voids and cracks in the material during moisture diffusion. The diffusion of moisture in composite is affected by fibers in a number of ways. Firstly, the geometry of the diffusion path is dependent on the fiber volume [*26*,*27*]. Secondly, the interface has different diffusion characteristics depending upon the degree of adhesion [*15*]. In SMC composites, the fiber arrangement being random-in-plane, the diffusing water molecules take a tortuous path. In a composite containing a high fiber volume like in SMC-65, favorable sites like voids due to imperfect bonding exist, where diffusing water molecules can settle down. Such "bound" water molecules do not participate in further diffusion [*5*,*6*]. The bound water upon expansion at an elevated temperature pushes the material, creating osmotic pressure and showing up on the surface as blisters. When blisters rupture, the interior material leaches out gradually.

The zinc stearate added to polyester resin to facilitate mold release migrates to the surface of the SMC laminate at the time of manufacture [*28*]. Zinc stearate, being inorganic material, does not react with ester molecules but forms a thin layer on the surface [*28*]. This fact is, however, in dispute [*29*]. Nevertheless,

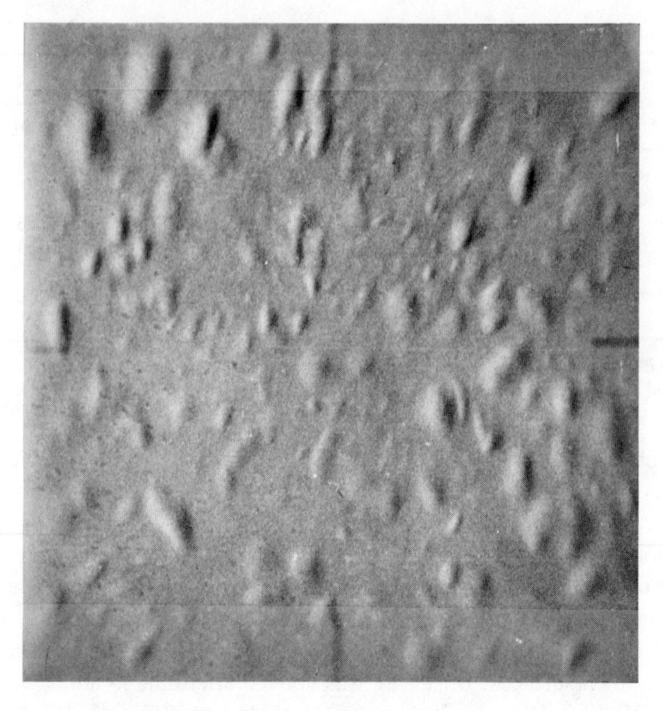

FIG. 4—*Surface damage in SMC-R65 after 100 days of exposure at 75°C in water (×100).*

the soft surface layer is more amenable to deformation than the stiff core. Furthermore, the compressive stress induced in the surface layer due to the expansional mismatch causes the debonded layer to buckle, thereby helping the blister grow. A discussion on plasticization and swelling of polyester resins can be found in Refs *30* and *31*.

Recalling that blisters upon rupture release entrapped water, the chemical nature of that water was evaluated with a Fisher Accumen pH meter. Table 5 lists the pH values of salt solutions and distilled water used in this study both before and after the sorption test. After the test, the distilled water had turned a little basic due to the formation of a magnesium hydroxide base from the magnesium oxide used as the thickener in the specimens and the absorbed water [*28*]

$$2\left(\mathrm{R}-\overset{|}{\underset{|}{\mathrm{c}}}-\overset{\|}{\mathrm{c}}-\mathrm{OH}\right) + \mathrm{MgO}$$

$$\rightarrow \mathrm{R}-\overset{|}{\underset{|}{\mathrm{c}}}-\overset{\|}{\mathrm{c}}-\mathrm{O}-\mathrm{Mg}-\mathrm{O}-\overset{\overset{\mathrm{O}}{\|}}{\underset{|}{\mathrm{c}}}-\overset{|}{\underset{|}{\mathrm{c}}}-\mathrm{R} + \mathrm{H_2} \quad (5)$$

The reaction described by Eq 5 is a possible mechanism of the leaching out of thickener material.

The fluid extracted from blisters is quite acidic, indicating a chemical degradation occurring inside the composite. When the blisters burst, they exposed glass fibers. Thus, blistering is believed to involve both mechanical debonding of glass fibers and a chemical degradation at interface.

Internal damage in specimens was examined with both an optical and a scanning electron microscope (SEM). In general, no sign of damage could be seen with an optical microscope at magnifications up to ×200 except in the severest environments of 98% RH and water immersion at 75°C. At 75°C/98% RH, the cross sections turned brown from white while they turned gray in 75°C/water. Under both environments, cracks and voids were observed.

SEM micrographs of specimens conditioned at 75°C in water are shown in Fig. 5. The cross section of the virgin specimen is clean, indicating good quality of the specimen. The damage initiation after environmental exposure seems to be in the form of small pits in resin-rich regions (see Fig. 5*b*). More extensive matrix damage is seen in Fig. 6. The damage takes the form of material degradation rather than a crack. Thus, chemical deterioration of the polyester resin is suspected under the humid environments at 75°C.

Chemical analysis was carried out on the residue found in environmental chambers. The residue appeared to be white clusters of substances containing salt upon partial evaporation of solutions in environmental chambers. Analysis later revealed the presence of small amounts of zinc stearate also. Zinc stearate forming a mold release layer on SMC specimens must have disintegrated after blister rupture.

The measured in-plane strains during absorption and desorption tests are shown in Figs. 7 and 8, respectively. The in-plane strains were almost equal in two perpendicular directions, confirming the quasi-isotropic nature of SMC materials. Therefore, the combined data are shown in Fig. 7. Shrinkage during desorption traces back swelling during absorption, and thus swelling is quite reversible in spite of the damages discussed earlier.

Although the swelling strain is not exactly proportional to the weight change under each environment, the combined data suggest a linear approximation. Since the slope can be taken as the swelling coefficient, it is interesting to predict this slope based on the rule-of-mixture equations.

The constituent properties required in the analysis are listed in Table 6. Since fibers are long enough to be considered continuous [*32,33*], the equations for continuous fiber composites in Ref *32* were used to determine the unidirectional properties. In addition, the volume additivity was assumed to estimate the swelling coefficient of polyester. The predicted unidirectional composite properties are listed in Table 7.

Treating SMC panels as quasi-isotropic laminates, we can use laminated plate theory [*32*] to derive an equation for in-plane swelling coefficient as

$$\beta_I = \frac{\beta_L + (\beta_L + \beta_T)v_{LT}E_T/E_L + \beta_T E_T/E_L}{1 + (1 + 2v_{LT})E_T/E_L} \quad (6)$$

FIG. 5—*SEM micrographs of cross sections:* (a) *virgin specimen* ($\times 125$); (b) *small pits in matrix-rich zone* ($75°C$/*water exposure*) ($\times 1250$).

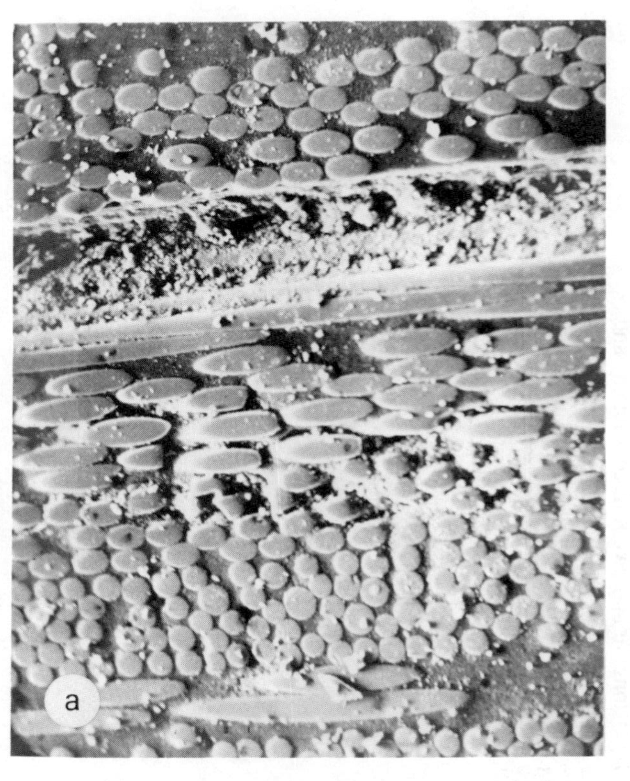

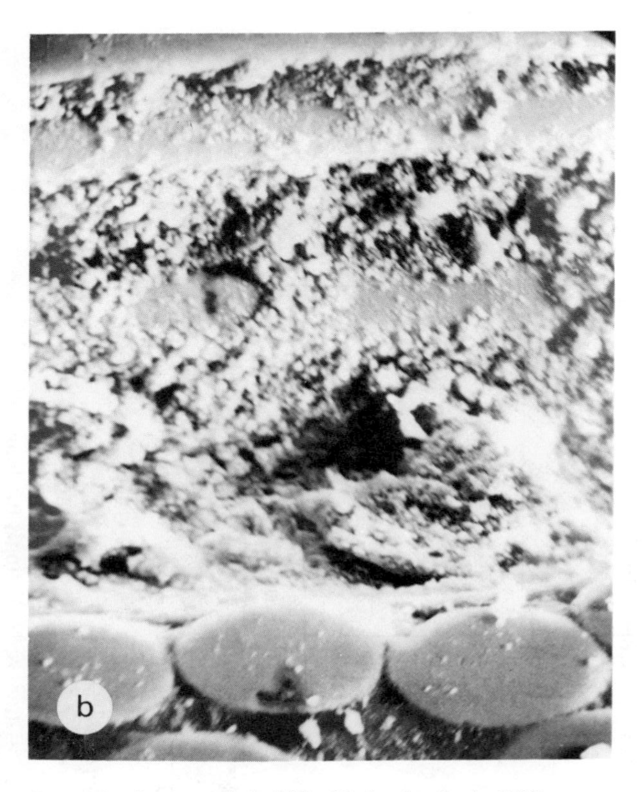

FIG. 6—*SEM micrographs of cross sections (75°C/water):* (a) *degradation of polyester resin (×250);* (b) *details of* a *(×1250).*

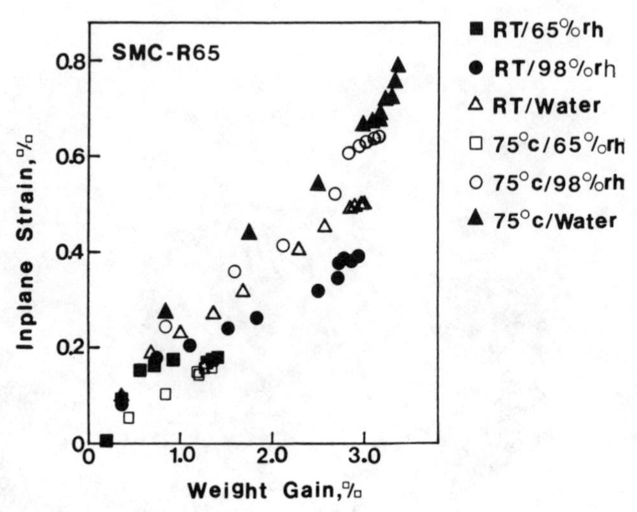

FIG. 7—*Variation of in-plane strain versus weight gain during moisture adsorption in SMC-R65.*

where

E_L = longitudinal Young's modulus,
E_T = transverse Young's modulus,
υ_{LT} = major Poisson's ratio,
β_L = longitudinal swelling coefficient, and
β_T = transverse swelling coefficient.

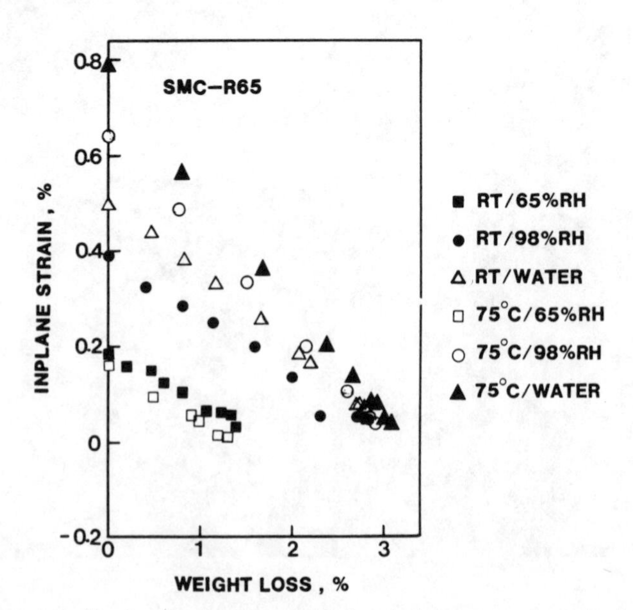

FIG. 8—*Variation of in-plane strain versus weight loss during moisture desorption in SMC-R65.*

TABLE 6—*Constituent properties.*

Property	E-Glass	Polyester
Young's modulus, GPa	72.35	2.80
Poisson's ratio	0.22	0.35
Density, 10^{-3} kg/m^3	2.60	1.24

TABLE 7—*Predicted unidirectional properties.*

Longitudinal Young's modulus	35.49 GPa
Transverse Young's modulus	7.01 GPa
Major Poisson's ratio	0.29
Longitudinal swelling coefficient	0.049
Transverse swelling coefficient	0.83

Substitution of the unidirectional properties into Eq 6 leads to $\beta_I = 0.20$. Despite several assumptions made in the analysis, the predicted swelling coefficient very well represents the average slope of the swelling data in Fig. 7.

In an earlier study on a different set of the same SMC specimens exposed to identical environments, ultrasonic attenuation measurements as well as short beam shear tests were conducted to correlate material degradation and mechanical property retention [23]. Manifestation of ultrasonic attenuation in materials due to hygrothermal degradation is shown in Fig. 9, while variation in shear

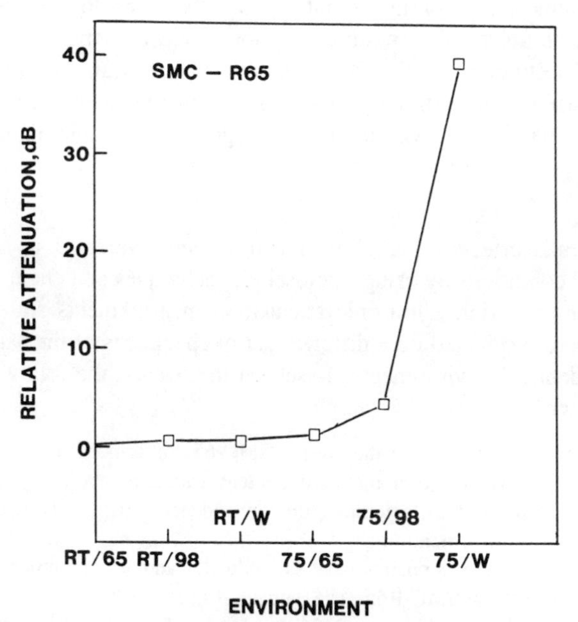

FIG. 9—*Effect of hygrothermal exposure on ultrasonic attenuation taken from Ref* 23.

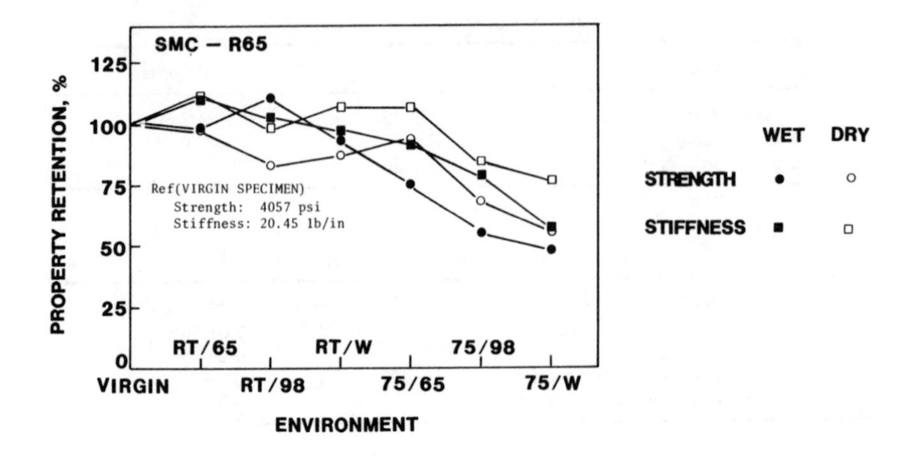

strength and stiffness after environmental aging can be seen in Fig. 10. Figures 9 and 10 were taken from Ref *23*. Here, relative attenuation is defined as 20 log A_0/A where A_0 and A are the magnitudes of the reflected pulse in the specimen which have been subjected to the reference environment (RT/65% RH) and the test environment, respectively. Relative attenuation increases with the severity of the environment showing progressive material deterioration. In Fig. 10, "wet" refers to specimens exposed to humid environments until saturation and later kept at laboratory atmosphere for three months. The "dry" refers to specimens that had undergone a complete absorption-desorption cycle and that were kept in a vacuum until testing. From the figure, it is seen that at all environments, wet specimens suffered more strength loss than the dry specimens. It is concluded, therefore, that part of the hygrothermal damage is recoverable upon drying.

Conclusions

Environmental effects on SMC-65 material were evaluated in six different hygrothermal conditions by using microscopic techniques and chemical analysis of leached out material. Ultrasonic attenuation measurements and short beam shear tests were conducted on a different set of specimens of the same material exposed to identical environments. Based on the results, the following conclusions are drawn:

1. The diffusion behavior of the present SMC-65 conformed with previously published data. Equilibrium moisture content varied parabolically with relative humidity and had little temperature dependence. Diffusivity decreased with increasing relative humidity.
2. Highly humid environments, that is, 98% RH and water immersion, at 75°C caused damage in SMC-R65. The damage manifested itself in (1) color changes internally and on the surfaces, (2) blistering on surfaces, (3) chemical degradation

of the matrix resin, (4) an increase in ultrasonic attenuation, and (5) reduced short beam shear strength.

3. At room temperature environments, there was no visual sign of permanent damage. Yet, the short beam shear strength was reduced by as much as 10% due to the presence of moisture in the interfaces.
4. Blisters involve degradation of the fiber-matrix interface, exposing bare fibers when burst. The liquid contained in blisters was acidic.
5. The internal damage is more chemical than mechanical because profuse deterioration, instead of sharp cracking, of the resin was observed on cross sections.
6. In-plane swelling of SMC-65 can be predicted from the rule-of-mixtures equation for continuous fiber composites and the volume additivity for swelling of resin.
7. Ultrasonic attenuation increased with the extent of damage. Severe damage at 75°C/98% RH and 75°C/water environments was unequivocally detected by ultrasonic attenuation.
8. Partial shear strength recovery can be realized upon removal of moisture.

References

[1] Shirrel, C. D. and Halpin, J. in *Composite Materials: Testing and Design (Fourth Conference), ASTM STP 617,* American Society for Testing and Materials, Philadelphia, pp. 514–528.

[2] Adamson, M. J., *Journal of Materials Science,* Vol. 15, 1980, pp. 1736–1745.

[3] Kelley, F. N. and Bueche, F., *Journal of Polymer Science,* Vol. 50, 1961, p. 549.

[4] Fox, T. G. and Flory, R. J., *Journal of Applied Physics,* Vol. 21, 1950, p. 581.

[5] Gurtin, M. E. and Yatomi, C., *Journal of Composite Materials,* Vol. 13, April 1979, pp. 126–130.

[6] Carter, H. C. and Kibler, K. G., *Journal of Composite Materials,* Vol. 12, 1978, pp. 118–131.

[7] Crank, J. and Park, G. S., Eds., *Diffusion in Polymers,* Chapters 5 and 8, Academic Press, London, 1968.

[8] Breuggemann, W. H., "Blistering of Gel Coat Laminates," 34th Annual Conference, Society of Plastics Industry, Section 4-E, Jan. 1979.

[9] Adams, R. C., "Variables Influencing the Blister of Marine Laminates," 37th Annual Conference, Society of Plastics Industry, Section 21-B, Jan. 1982.

[10] Ghotra, J. S. and Pritchard, G., 28th National Symposium, Society for the Advancement of Material and Process Engineering, April 1983, pp. 807–817.

[11] Edwards, D. B. and Sridharan, N. S., "Environmental Testing of High Strength Molding Compounds," an unpublished report prepared at Science and Technology Laboratory, International Harvester Co., Hinsdale, IL, 1979.

[12] Blaga, A. and Yamasaki, R. D., *Journal of Material Science,* No. 8, 1973, pp. 1131–1139.

[13] Ashbee, K. H. and Wyatt, R. C., *Proceedings of Royal Society,* A312, 1969, pp. 553–564.

[14] Ishai, O., *Polymer Engineering and Science,* Vol. 15, No. 7, July 1975, pp. 486–499.

[15] Bueche, F., *Physical Properties of Polymers,* Interscience Publishers, New York, 1962.

[16] Yamasaki, R. S., *Composite Technology Review,* Vol. 4, No. 3, Fall 1982, pp. 84–87.

[17] Sanders, B., Ed., *Short Fiber Reinforced Composite Materials, ASTM STP 772,* American Society for Testing and Materials, Philadelphia, April 1980.

[18] Kay, J. F., "Advances in Structural Sheet Molding Compound," Report 80–49, Research and Development Division, Owens-Corning Fiberglas Technical Center, Granville, OH, 1980.

[19] Anonymous, Process Engineering News, *Plastics Technology,* Sept. 1980, pp. 26–31.

[20] Lubin, George, Ed., *Handbook of Fiberglass and Advanced Plastics Composites,* Van Nostrand Reinhold, New York, 1969.

[21] Loos, A. C. and Springer, G. S., *Journal of Composite Materials,* Vol. 13, April 1979, pp. 131–146.

[22] Shirrel, C. D. and Duff, C. C., 29th National Symposium, Society for the Advancement of Material and Process Engineering, April 1983, pp. 893–894.

[*23*] Hahn, H. T. and Hosangadi, A., "Environmental Degradation of Sheet Molding Compounds," *American Chemical Society Organic Coatings and Plastics Chemistry,* Washington, DC, April 1983.

[*24*] Meares, P., *Polymers: Structure and Bulk Properties,* Van Nostrand, New York, 1965.

[*25*] Brinkley, S. R., *Principles of General Chemistry,* Macmillan, New York, 1944.

[*26*] Mehta, B. S., Dibenedetto, A. T., and Kardos, J. L., *International Journal of Polymeric Materials,* Vol. 5, 1976, pp. 147–161.

[*27*] Hull, D., *An Introduction to Composite Materials,* Cambridge University Press, New York, 1981.

[*28*] Bruins, P. E., Ed., *Unsaturated Polyester Technology,* Gordon and Breach Science Publishers, New York, 1976.

[*29*] Fletcher, C. W., "An Examination of the Mechanism by which Stearates Act as Internal Mold Release Agents in Matched Die RP Moldings," 37th Annual Conference, Society of Plastics Industry, Section 22-F, Jan. 1982.

[*30*] McCabe, M. V., "The Effects of Various Chemical Environments on the Flexural Properties of Molded SMC," 34th Annual Conference, Society of Plastics Industry, Section 22-C, 1979.

[*31*] Edwards, H. R., "Variables Influencing the Performance of a Gel Coated Laminate," 34th Annual Conference, Society of Plastics Industry, Section 4-D, 1979.

[*32*] Tsai, S. W. and Hahn, H. T., *Introduction to Composite Materials,* Technomic Publishing Co., Westport, CT, 1980.

[*33*] Taggart, D. G. et al, "Properties of SCM Composites," Report No. CCM-79-1, Center for Composite Materials, University of Delaware, Newark, DE, Feb. 1979.

Alfred C. Loos[1] *and William T. Freeman, Jr.*[2]

Resin Flow During Autoclave Cure of Graphite-Epoxy Composites

REFERENCE: Loos, A. C. and Freeman, W. T., Jr., **"Resin Flow During Autoclave Cure of Graphite-Epoxy Composites,"** *High Modulus Fiber Composites in Ground Transportation and High Volume Applications, ASTM STP 873,* D. W. Wilson, Ed., American Society for Testing and Materials, Philadelphia, 1985, pp. 119–130.

ABSTRACT: Resin flow during autoclave cure of graphite-epoxy composites was measured experimentally as a function of time. The data indicate that the ply-stacking sequence of the composite does not significantly influence the total resin mass loss. Resin flow data obtained from composites with different ply thicknesses and dimensions were compared with the results of the model of Loos and Springer. Good agreement was found between the data and the model.

KEY WORDS: autoclave, composites, cure, graphite-epoxy composites, resin flow

Composite structures constructed from graphite-epoxy prepreg materials are fabricated by laminating multiple plies into the desired shape and then curing the material in an autoclave by simultaneous application of heat and pressure. Elevated temperature applied during the cure causes resin polymerization and cross-linking while the applied pressure consolidates the individual prepreg plies by squeezing out excess resin. The magnitude and duration of cure cycle temperatures and pressures significantly influence the final physical and mechanical properties of the composite. A composite that is processed using an optimum cure cycle will result in a void-free structure that is uniformly cured to the desired resin content and glass transition temperature in the shortest amount of time. Therefore, the cure cycle must be carefully selected for each application considering the composition of the prepreg material and the geometry of the structure.

As a result of previous studies, the effect of temperature on the curing process is well understood [*1–9*]. The analytical techniques presented in these in-

[1]Assistant professor, Department of Engineering Science and Mechanics, Virginia Polytechnic Institute and State University, Blacksburg, VA 24061.

[2]Aerospace technologist, Polymeric Materials Branch, NASA-Langley Research Center, Hampton, VA 23665.

vestigations can be used to predict the temperature distribution, degree of cure, and cure time of graphite-epoxy composites.

However, the influence of the cure cycle, the prepreg properties, and the composite ply stacking sequence on the resin flow process occurring during cure has not been studied extensively. Analytical models that can determine the effects of these various processing parameters on the resin flow process have been only recently proposed [*9–11*]. Resin flow measurements which can be used to assess the accuracy of the models are rather limited. Only one recent study reported the effects of cure pressure, composite thickness, and prepreg resin content on the resin flow process of small unidirectional graphite-epoxy panels [*9*]. The influence of the dimensions and the ply stacking sequence of the composite on the resin flow process has not been examined.

In this paper experimental results are presented which illustrate the effects of composite geometry on the resin flow process of autoclave cured graphite-epoxy composites. In addition, the resin flow data obtained are compared to the results of the cure process model presented by Loos and Springer [*9*].

Experimental Procedure

Composite specimens for this program were fabricated from a 32-kg (70-lb) roll of Hercules AS/3501-6 graphite-epoxy prepreg tape. The resin content of the prepreg was determined in accordance with the ASTM Method for Resin Content of Carbon and Graphite Prepregs by Solvent Extraction (C 613-67 [1980]) and found to be 43% by weight at the beginning of the roll and 42% by weight at approximately the middle of the roll. The prepreg had a fiber areal weight of 152 g/m^2 as quoted in the manufacturer's quality assurance certification.

Flat-plate, unidirectional, cross-ply, angle-ply, and quasi-isotropic lay-ups were fabricated to the dimensions given in Table 1. Upon completion of each lay-up, the specimen was precompacted under a vacuum of 635-mm (25-in.) mercury for 20 min at room temperature. All completed lay-ups were stored in sealed polyethylene bags at −18°C (0°F).

Resin flows, both normal to the plane of the composite and in the plane of the composite, were measured during cure by using a bleeder mounted on the top of

TABLE 1—*Ply stacking sequences and dimensions of the test specimens.*

Ply Stacking Sequence	Number of Plies	Dimensions, Length by Width
$[0]_{32}$	32	15.2 by 15.2 cm
$[0/90]_{8s}$	32	15.2 by 15.2 cm
$[+45/-45]_{8s}$	32	15.2 by 15.2 cm
$[0/+45/90/-45]_{4s}$	32	15.2 by 15.2 cm
$[0/90]_{4s}$	16	15.2 by 15.2 cm
$[0/90]_{16s}$	64	15.2 by 15.2 cm
$[0/90]_{8s}$	32	7.6 by 7.6 cm
$[0/90]_{8s}$	32	30.5 by 30.5 cm

the lay-up (vertical bleedout) and bleeders positioned around the periphery of the lay-up (edge bleedout). The vertical or top bleeder was constructed from Mochburg CW 1850 bleeder material, cut to the length and width of the composite lay-up. The required thickness of the top bleeder depends on the ply thickness of the composite and the amount of resin to be bled out. For the present study, 5 sheets of Mochburg were used to bleed a 16-ply lay-up, 11 sheets of Mochburg were used to bleed a 32-ply lay-up, and 21 sheets of Mochburg were used to bleed a 64-ply lay-up. This gave a prepreg to bleeder ratio of about three to one.

Edge bleeders were fabricated from narrow strips of Air Weave N-10 bleeder cloth, carefully cut to have the same dimensions as the edges of the lay-up. The thicknesses of the edge bleeders varied with the ply thickness of the lay-up. The 16, 32, and 64-ply lay-ups required edge bleeders made of one, three, or four strips of Air Weave N-10, respectively. One edge bleeder was placed along each of the four edges of the lay-up, except for the unidirectional lay-ups. Edge bleeders parallel to the fibers of the unidirectional lay-ups were not used to prevent excessive fiber spread caused by resin flow in the plane of the composite, normal to the fibers.

Before each test, the lay-ups, top bleeders, and edge bleeders were weighed on either a Mettler 120-g capacity or a Mettler 1200-g capacity electronic pan balance. The eight lay-ups shown in Table 1 plus two additional 15.2 by 15.2 cm, 32-ply, cross-ply lay-ups with special bleeder configurations were mounted on a single 1.2-m-wide by 2.4-m-long by 1.27-cm-thick tool plate. In order to measure the actual cure temperature, one No. 30 gage iron-constantan thermocouple was attached to the top and bottom surfaces of each lay-up. The bleeders were positioned around each lay-up and assembled according to the schematic shown in Fig. 1. A complete list of all materials used in the bagging procedure is given in Table 2. The composite-bleeder assemblies were covered with a single Capran

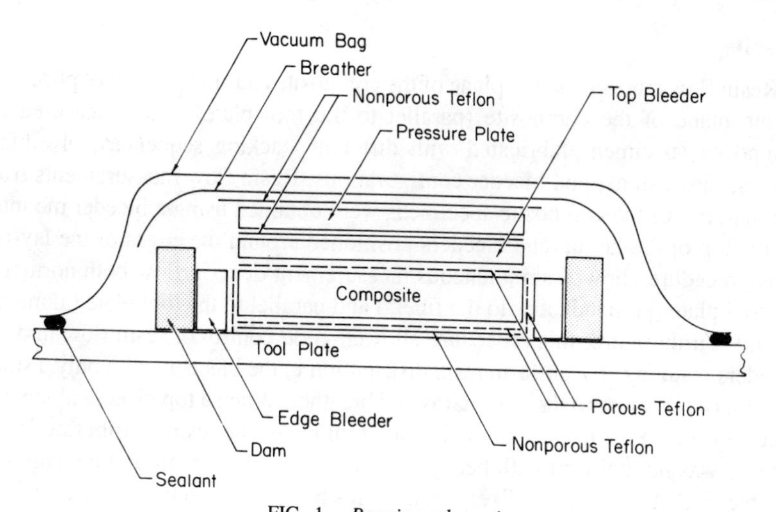

FIG. 1—*Bagging schematic.*

TABLE 2—*Materials used in the bagging assembly (see Fig. 1).*

Item	Material
Nonporous Teflon	Air Tech 234 TFNP
Porous Teflon	Air Tech 234 TFP
Top bleeder	Mochburg CW 1850
Edge bleeders	Air Tech Air Weave N-10
Pressure plate	1/8-in.-thick mild steel
Breather	Air Tech Air Weave N-10
Vacuum bag	Capran-80 polyamide film
Dams	Corprene DK-153
Sealant	General Sealants GS-43

vacuum bag. The bag was connected to a vacuum pump and a full vacuum of 734-mm (29 in.) mercury applied. The tool plate was placed in a Devine bonding autoclave and the autoclave was pressurized to 345 kPa (50 psi). Both the vacuum bag pressure and the autoclave pressure remained constant for the duration of the cure.

The autoclave was heated at a rate of 2.8°C/min (5°F/min) from room temperature to the desired cure temperature of 116°C (240°F). During cure the autoclave air temperature and pressure, the vacuum bag pressure, and the temperatures at the top and bottom surfaces of each lay-up were continuously monitored by a computer-controlled data acquisition system.

At preselected times, the autoclave heaters were turned off, the pressure was released, and the vacuum pump was shut off. After the composite specimens were cooled to room temperature, each composite-bleeder assembly was carefully removed from the tool plate and the weights of the composite, top, and edge bleeders were recorded.

Results

Resin flows, normal to the plane of the composite (normal to the tool plate) and in the plane of the composite (parallel to the tool plate), were measured for composite specimens fabricated with different stacking sequences, ply thicknesses, dimensions, and bleeder configurations. Resin flow measurements from the majority of the composite specimens were obtained using a bleeder mounted on the top of the lay-up and bleeders positioned around the edges of the lay-up. This procedure allowed simultaneous measurement of resin flow both normal to the tool plate (perpendicular to the fibers) and parallel to the tool plate (along the fibers). Furthermore, this procedure allowed good control of resin flow into the bleeders resulting in a uniform resin distribution in the bleeders with only a small amount of resin loss to the edge dams and breather. When a top bleeder alone was used to measure resin flow normal to the tool plate, the resin distribution in the bleeder was nonuniform with heavy concentrations of resin around the edges of the bleeder. This is most likely due to resin leakage from the edges of the composite into the top bleeder. Similar problems with nonuniform resin distributions in the bleeders and leakage of resin into the edge dams and breather were observed when edge bleeders alone were used to measure resin flow in the plane of the composite. It should be noted that the measured resin flow data reported in this study were obtained by assuming that the measured resin in the top bleeder is due entirely to resin flow from the top of the composite, normal to the tool plate, and the measured resin in the edge bleeders is due entirely to resin flow from the edges of the composite, parallel to the tool plate.

The same cure cycle was used in every test and is shown in Fig. 2. The cure temperature was measured by thermocouples attached to the top and bottom surfaces of each composite lay-up. During cure the maximum difference in temperature between any two composite lay-ups was less than 10°C and the largest variation in temperature between the upper and lower boundaries of any one composite was less than 5°C. The temperature curve shown in Fig. 2 represents the average temperature of the top and bottom surfaces of each composite.

The results of the resin flow measurements are shown in Figs. 3–7. In these figures, time t is plotted on the abscissa. The ordinates represent either the total resin mass loss m^* of the composite or the resin mass losses due to flow in the directions normal (m_T^*) and parallel (m_E^*) to the tool plate in time t. The mass losses shown in Figs. 3–7 represent the mass loss with respect to the initial mass of the composite

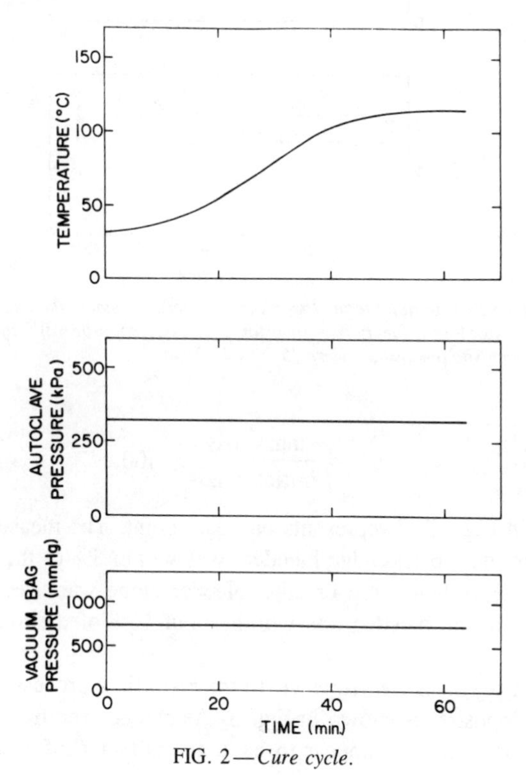

FIG. 2—*Cure cycle.*

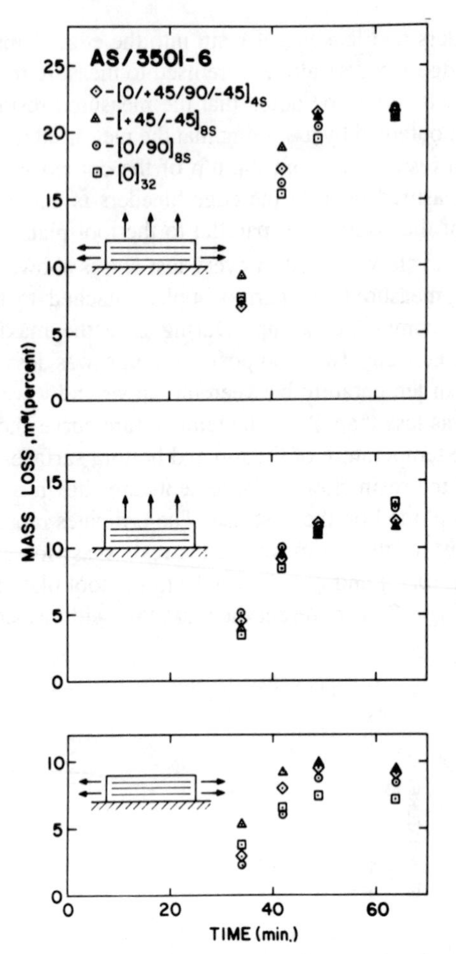

FIG. 3—*The mass loss, parallel to the tool plate* (bottom), *normal to the tool plate* (center), *and the total mass loss* (top) *as a function of time for a 32-ply composite with different ply stacking sequences. The composite dimensions were 15.2 by 15.2 cm.*

$$m^* = \frac{\text{mass loss}}{\text{initial mass}} \times 100$$

Each symbol in Figs. 3–7 represents one data point. The measured sum of the weight gains in the top and edge bleeders was within 5% of the total mass loss of the composite. When a top or edge bleeder alone was used, the measured resin contained in the bleeder was within about 12% of the total mass loss of the composite.

The effect of ply stacking sequence on the resin flow process for 32-ply, 15.2 by 15.2 cm composites is shown in Fig. 3. As can be seen from the figure, the stacking sequence does not appear to have a significant influence on the resin

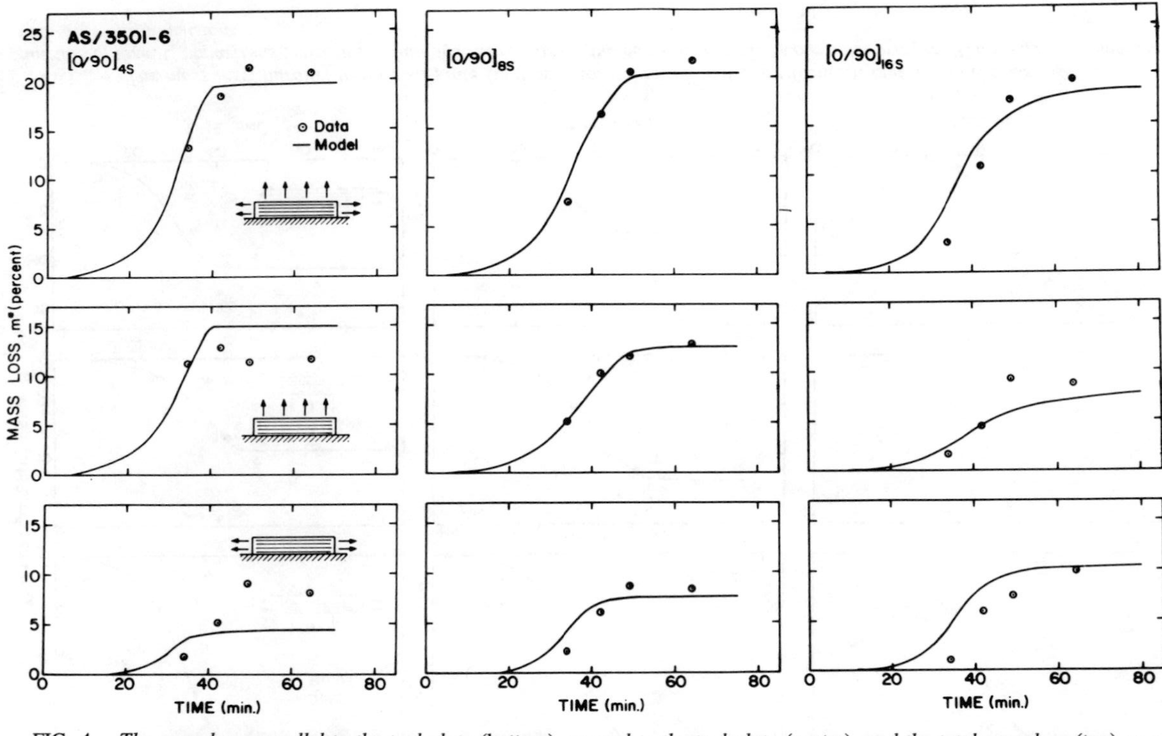

FIG. 4—*The mass loss, parallel to the tool plate* (bottom), *normal to the tool plate* (center), *and the total mass loss* (top) *as a function of time. Comparisons between the data and the results computed by the model of Loos and Springer* [9] *for 16, 32, and 64-ply composites. The composite dimensions were 15.2 by 15.2 cm.*

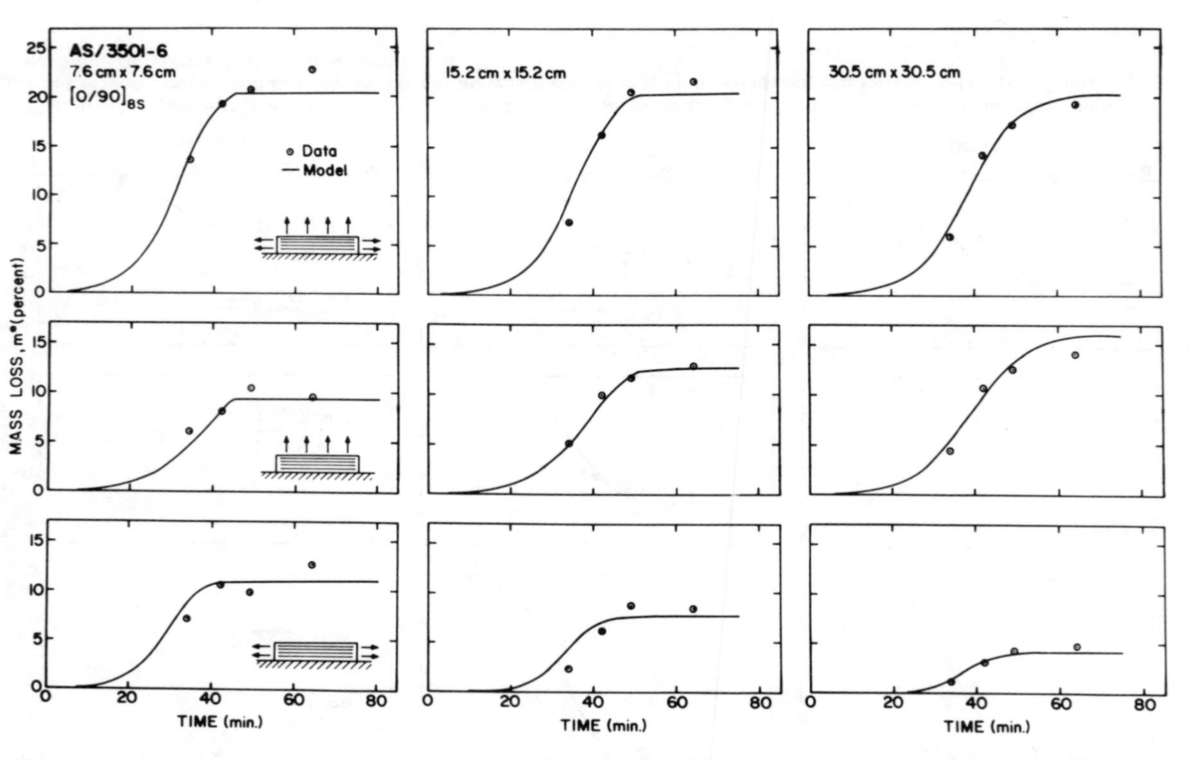

FIG. 5—*The mass loss, parallel to the tool plate* (bottom), *normal to the tool plate* (center), *and the total mass loss* (top) *as a function of time. Comparisons between the data and the results computed by the model of Loos and Springer* [9] *for a 32-ply composite with different dimensions.*

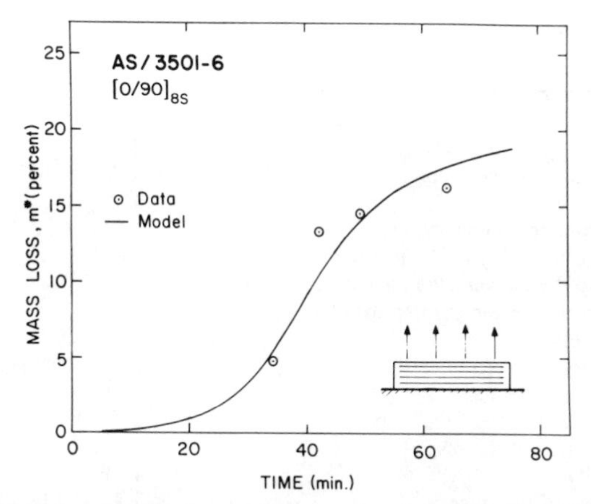

FIG. 6—*The mass loss normal to the tool plate as a function of time for a 32-ply composite using a top bleeder only. Comparisons between the data and the results computed by the model of Loos and Springer* [9]. *The composite dimensions were 15.2 by 15.2 cm.*

flow normal to the tool plate. The differences between the total resin loss from composites with the four stacking sequences is primarily due to variations in the amount of resin flow from the edges of the composite, parallel to the tool plate. However, near the completion of the resin flow process, the total resin loss is the same for all composites regardless of the stacking sequence employed.

At a cure time of 49 min the measured resin flow from the edges of the composite, parallel to the tool plate, actually exceeds the edge flow at a later time of 64 min. This result was observed only in the edge flow measurements and may have been caused by a slight variation in the cure cycle. In view of these results, additional experiments will have to be performed to determine the accuracy of the

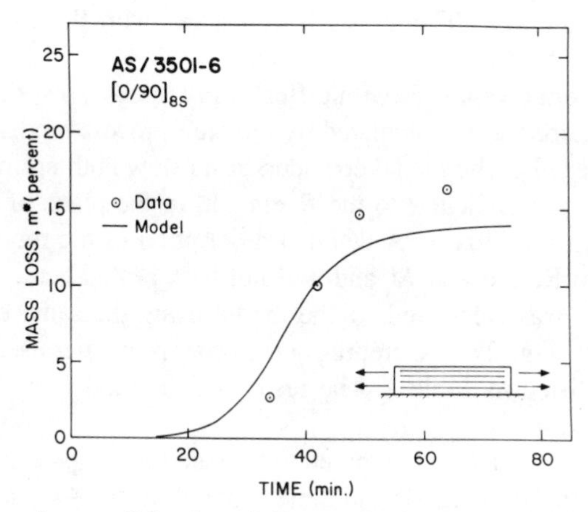

FIG. 7—*The mass loss parallel to the tool plate as a function of time for a 32-ply composite using edge bleeders only. Comparisons between the data and the results computed by the model of Loos and Springer* [9]. *The composite dimensions were 15.2 by 15.2 cm.*

TABLE 3—*Properties of Hercules AS/3501-6 prepreg and Mochburg CW 1850 thermal fiber bleeder cloth.*

AS/3501-6	
Initial prepreg resin mass fraction	43%
Initial thickness of the prepreg	1.778×10^{-4} m
Fiber areal weight	152 g/m^2
Thickness of one compacted ply	1.321×10^{-4} m
Apparent permeability of the prepreg normal to the plane of the composite	2.0×10^{-15} m^2
Flow coefficient of the prepreg parallel to the fibers	1.7×10^2
Resin density	1.26×10^3 kg/m^3
Specific heat of the resin	1.26 kJ/(kg · K)
Thermal conductivity of the resin	1.67×10^{-1} W/(m · K)
Heat of reaction of the resin	474 J/g
Fiber density	1.79×10^3 kg/m^3
Specific heat of the fiber	7.12×10^{-1} kJ/(kg · K)
Thermal conductivity of the fiber	2.60×10^1 W/(m · K)
Relationship between the cure rate, temperature, and degree of cure	see Ref *13*
Relationship between viscosity, temperature, and degree of cure	see Ref *13*
MOCHBURG BLEEDER CLOTH	
Apparent permeability	5.6×10^{-11} m^2
Porosity	0.57

resin flow measurements and verify the stacking sequence effect on the resin flow in the plane of the composite, parallel to the tool plate.

The effects of the composite ply thickness and the composite dimensions on the resin flow process are shown in Fig. 4 and Fig. 5, respectively. The data indicate that as the length-to-thickness ratio of the composite increases the ratio of resin flow normal to the plane of the composite to resin flow in the plane of the composite also increases. This result is consistent with the data presented in Ref *9*.

Resin flows, from composites with different ply thicknesses, dimensions, and bleeder configurations were calculated by the cure process model developed by Loos and Springer [*9*]. The model considers resin flow both normal to the plane of the composite perpendicular to the fibers and in the plane of the composite along or parallel to the fibers. A detailed description of the model has already been presented in Refs *9* and *12* and will not be repeated here.

The resin flow was calculated by the model using the same cure cycle employed in the tests (Fig. 2). The prepreg and bleeder properties used in the present calculations are listed in Table 3. The results of the model are represented by solid lines in Figs. 4–7.

The results of the model are compared to the data for composite laminates with different ply thicknesses (Fig. 4) and dimensions (Fig. 5). As can be seen, there is good agreement between the calculated and measured resin mass loss. The assumption that the resin collected in the top bleeder is due entirely to resin flow normal to the tool plate and that the resin collected in the edge bleeders is due entirely to resin flow parallel to the tool plate appears reasonable.

Only in the case of the 16-ply, $[0/90]_{4s}$ laminate did the predicted resin mass loss not agree with the measured resin mass loss. Due to the high resin content of the prepreg, the top bleeder became fully saturated with resin during the initial portion of the cure cycle. Thus, for resin flow normal to the tool plate, the measured resin flow was considerably less than the calculated resin flow. The excess resin remaining in the composite could only flow out the edges of the composite resulting in a larger amount of resin flow parallel to the tool plate (Fig. 4).

Resin flow data from 32-ply composite laminates using either a top bleeder only or edge bleeders only are presented in Fig. 6 and Fig. 7, respectively. The agreement between the results of the resin flow model and the data is still reasonably good in view of the previously mentioned problems with uncontrolled resin losses into the edge dams and breather.

Conclusions

The following conclusions can be made based on the present data obtained with Hercules AS/3501-6 graphite-epoxy composites:

1. The composite ply stacking sequence does not have a significant effect on the resin flow normal to the plane of the composite, perpendicular to the fibers. The total resin mass loss of composites with different stacking sequences is nearly the same at the completion of the cure cycle.

2. Additional experiments must be performed to determine the accuracy of the resin flow measurements and verify the apparent stacking sequence effect on resin flow in the plane of the composite, parallel to the tool plate.

3. As the length-to-thickness ratio of the composite is increased the ratio of resin flow normal to the tool plate to resin flow parallel to the tool plate also increases.

4. The cure process model presented by Loos and Springer can be used to predict the resin flow from autoclave cured composites.

Acknowledgments

The experimental portions of the work were completed while the author (A. C. Loos) was a participant in the 1983 NASA/ASEE Summer Faculty Fellowship Program at NASA-Langley Research Center. The financial support received from the NASA-Virginia Tech Composites Program, NASA CA NCC1015 for the preparation of this paper is gratefully acknowledged. The authors wish to thank Frank Reaves, Keith Benson, and William Conkling for their assistance with the fabrication and processing of the composite specimens.

References

[1] Progelhof, R. C. and Throne, J. L., *Polymer Engineering and Science*, Vol. 15, No. 9, Sept. 1975, pp. 690–695.

[2] Broyer, E. and Macosko, C. W., *Journal,* American Institute of Chemical Engineers, Vol. 22, No. 2, March 1976, pp. 268–276.
[3] Adabbo, H. E., Rojas, A. J., and Williams, R. J. J., *Polymer Engineering and Science,* Vol. 19, No. 12, Sept. 1979, pp. 835–840.
[4] Stolin, A. M., Merzhanov, A. G., and Malkin, A. Y., *Polymer Engineering and Science,* Vol. 19, No. 15, Nov. 1979, pp. 1065–1073.
[5] Barone, M. R. and Caulk, D. A., *International Journal of Heat and Mass Transfer,* Vol. 22, No. 7, July 1979, pp. 1021–1032.
[6] Pusatcioglu, S. Y., Hassler, J. C., Fricke, A. L., and McGee, H. A., Jr., *Journal of Applied Polymer Science,* Vol. 25, No. 3, March 1980, pp. 381–393.
[7] Lee, L. J., *Polymer Engineering and Science,* Vol. 21, No. 8, June 1981, pp. 483–492.
[8] Nied, H. A., *Technical Papers,* Vol. 27, Society of Plastics Engineers, Greenwich, CT, 1981, pp. 344–347.
[9] Loos, A. C. and Springer, G. S., *Journal of Composite Materials,* Vol. 17, No. 2, March 1983, pp. 135–169.
[10] Lindt, J. T., *Quarterly,* Society for the Advancement of Material and Process Engineering, Vol. 14, No. 1, Oct. 1982, pp. 14–19.
[11] Bartlett, C. J., *Technical Papers,* Vol. 24, Society of Plastics Engineers, Greenwich, CT, 1978, pp. 638–640.
[12] Loos, A. C. and Springer, G. S., "Curing of Graphite/Epoxy Composites," Technical Report AFWAL-TR-83-4040, Wright Aeronautical Laboratories, Wright Patterson Air Force Base, Dayton, OH, 1983.
[13] Lee, W. I., Loos, A. C., and Springer, G. S., *Journal of Composite Materials,* Vol. 16, No. 6, Nov. 1982, pp. 510–520.

Shing-Chung Yen, [1] *Clement Hiel,* [2] *and Donald H. Morris* [3]

Viscoelastic Response of SMC-R50 Under Different Thermomechanical Conditions

REFERENCE: Yen, S. C., Hiel, C., and Morris, D. H., **"Viscoelastic Response of SMC-R50 Under Different Thermomechanical Conditions,"** *High Modulus Fiber Composites in Ground Transportation and High Volume Applications, ASTM STP 873,* D. W. Wilson, Ed., American Society for Testing and Materials, Philadelphia, 1985, pp. 131–143.

ABSTRACT: The creep and creep recovery behavior of a random fiber composite (SMC-R50) was investigated. Time-dependent behavior at various stress and temperature levels was experimentally determined. Findley's equation was used to model the viscoelastic behavior of short-term tests. The Findley equation was also used to predict the time-dependent behavior of the random fiber composite for a multiple step loading. Excellent agreement between experimental results and the Findley equation was obtained. Since repeatable results were obtained only for mechanically conditioned specimens, the test results were compared to experimental data obtained from unconditioned specimens. Experimental results of the conditioned specimens fell within the scatter band of the data for the unconditioned specimens.

KEYWORDS: short fiber composite, viscoelastic behavior, creep of composites, sheet molding compound, power law creep

Since the early 1950s fiber-reinforced composites have received considerable attention by many investigators. Composite materials are valued not only because of their light weight but also because of their durability, dimensional stability, and favorable chemical, mechanical, and electrical properties [*1*]. Of many composite materials, sheet molding compound (SMC) is one of the candidate materials that is being considered to replace some of the metallic structural members in automobiles [*2*]. SMC is a composite material made of glass fibers embedded in a polyester matrix [*3*]. The composite molding process offers design flexibility

[1]Assistant professor, Department of Civil Engineering and Mechanics, Southern Illinois University, Carbondale, IL 62901.

[2]Research associate, NASA Ames Research Center, Moffett Field, CA 94035.

[3]Professor, Department of Engineering Science and Mechanics, Virginia Polytechnic Institute and State University, Blacksburg, VA 24061.

and allows part consolidation, ultimately reducing the installed cost of complex components [*4*].

If SMC is to find wider usage in the automobile industry, it is important that its mechanical properties be understood and, where possible, optimized [*2*]. Currently SMC has achieved limited automotive usage in load-carrying members due to the lack of understanding of some aspects of its mechanical behavior. In particular, it is known that SMC exhibits viscoelastic or time-dependent effects. This time-dependent behavior is significantly affected by stress, temperature, humidity, etc. [*5,6*]. Because of these effects, there is concern that time-dependent phenomena such as creep and creep rupture may be of importance in long-term design.

A limited amount of research into the viscoelastic behavior of SMC has been published. Longtime creep response of various SMC material systems has been experimentally studied by the researchers at General Motors [*1,2,4*] and Owens-Corning Co. [*3*]. Their data exhibited considerable scatter and sudden jumps in creep strain. In another report, Cartner, Griffith, and Brinson [*7*] used a repeated loading and unloading procedure to mechanically condition SMC-R25 specimens prior to creep and creep recovery tests. The mechanical conditioning eliminated the jumps in the creep strain and gave repeatable creep and recovery responses from test to test. They were then able to describe the creep behavior of SMC-R25 using the Schapery procedure [*8*]. More recently, Jerina et al [*9*] developed a mechanical conditioning procedure to obtain repeatable creep and recovery response of SMC-R50. Furthermore, Ref *9* indicated that as long as the stress level used in such a mechanical conditioning procedure was not too high, a nearly constant flaw state could be reached and the basic creep characteristics preserved.

In this paper, the effects of both stress and temperature on the uniaxial creep of conditioned SMC-R50 specimens are presented. Short-time creep results were modelled by the Findley equation [*10*]; this equation was used to predict the longtime creep response as well as the response due to a steploading. To differentiate the creep behavior of conditioned and unconditioned SMC-R50 specimens, comparisons are made between the creep response of conditioned and unconditioned [*3*] specimens.

Mathematical Model

The analytical expression used to model the creep and recovery response of the SMC-R50 material was the Findley equation [*10*], which may be written

$$\varepsilon(t) = \varepsilon_0 + mt^n \tag{1}$$

where

$\varepsilon(t)$ = time-dependent creep strain,
ε_0 = instantaneous strain,
t = time, and
m and n = experimentally determined material parameters.

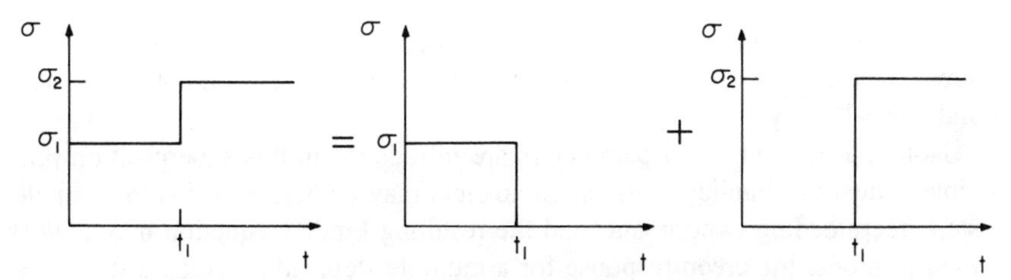

FIG. 1—*Illustration of modified superposition principle.*

The parameters ε_0 and m may be functions of stress, temperature, moisture, etc., whereas n may be a function of temperature, moisture, etc. but is independent of stress.

Furthermore, Findley [*10–12*] used the following relationships for ε_0 and m

$$\varepsilon_0 = \varepsilon_0' \sinh(\sigma/\sigma_e) \tag{2}$$

$$m = m' \sinh(\sigma/\sigma_m) \tag{3}$$

where ε_0', m', σ_e, and σ_m are material parameters that may be functions of temperature, moisture, etc. but are independent of the applied uniaxial stress, σ. The Findley equation may be used to model both the creep and recovery response of a uniaxial specimen subjected to a constant load.

A structural member during service may be frequently subjected to abrupt changes of loading due to the failure of other members or due to a sudden load increase. To account for this kind of phenomenon, Findley and Lai [*13*] developed a modified superposition principle. For uniaxial tension, this superposition method can be described as follows [*13*]: when the state of stress is abruptly changed from σ_1 to σ_2 at time t_1, the creep behavior can be considered as if at this instant stress σ_1 is removed and at the same time stress σ_2 is applied to the specimen, both being considered as independent actions. To illustrate this method, consider a two-step loading as shown in Fig. 1. The recovery strain ε' resulting from removal of σ_1 after loading time t_1 is given by

$$\varepsilon' = m(\sigma_1)[t^n - (t - t_1)^n] \tag{4}$$

The creep behavior due to σ_2 applied at t_1 denoted by ε'' is

$$\varepsilon'' = \varepsilon_0(\sigma_2) + m(\sigma_2)(t - t_1)^n \tag{5}$$

The total strain will be the sum of Eqs 4 and 5 and equal to

$$\varepsilon = \varepsilon_0(\sigma_2) + m(\sigma_1)[t^n - (t - t_1)^n] + m(\sigma_2)(t - t_1)^n \tag{6}$$

The strain response due to the stress changing k times is [*13*]

$$\varepsilon = \varepsilon_0(\sigma_k) + \sum_{p=1}^{k} [m(\sigma_p) - m(\sigma_{p-1})](t - t_{p-1})^n \tag{7}$$

In this theory, the time-independent part depends only on the current state of stress, whereas the time-dependent part of the strain depends on both stress state and stress history.

Note that no additional parameters are introduced in this superposition principle. Thus, the Findley material parameters may be determined from a single-step creep-loading experiment, and the resulting Findley equation may then be used to model the creep response for a multiple steploading creep test.

In addition, Findley and Lai [*13*] show that Eq 7 gives residual strains that approach zero for sufficiently long recovery times. This means that a specimen can be repeatedly used if the recovery time is such that nearly full recovery is allowed and provided that the material behaves as a purely viscoelastic solid [*13*].

Experimental Procedures

The Material and Specimen Configuration

The material used for the current work was SMC-R50, a SMC containing 50% by weight chopped glass fibers randomly oriented in the plane of the sheet [*3*]. It has been found [*3*] that SMC-R50 can be classified as an isotropic material, where the plane of isotropy coincides with the plane of the sheet.

The dog bone specimens used for testing were sawed from SMC-R50 panels with a diamond abrasive disk; the final shape was formed using a high-speed router. The length of the specimen was 203 mm (8 in.), while the gage section was 51 mm (2 in.) long and 13 mm (0.5 in.) wide. Specimen thickness was approximately 3 mm (0.125 in.).

Mechanical Conditioning

A study of the basic creep characteristics of SMC-R50 is complicated by sudden jumps that occur during creep testing [*1,3,4,9*]. To eliminate the jumps and to give repeatable creep responses, Cartner, Griffith, and Brinson [*7*] and Jerina, Schapery, Tung, and Sanders [*9*] found it necessary to mechanically condition specimens. The conditioning procedure adapted for the work reported herein may be found in Ref *9*. The procedure illustrated in Fig. 2 was used to mechanically condition specimens at each temperature. The conditioning stress at each temperature was 50% of the ultimate tensile strength at that temperature. In general, seven cycles were needed to obtain repeatable results. Figure 3 shows the strain response due to mechanical conditioning at room temperature.

Creep Experiments

Lever arm creep frames were used for loading the specimens. The frames were equipped with ovens that maintained temperatures within 1.1°C (±2°F) of the desired temperature. Creep strains were measured using strain gages. Gages were mounted on opposite sides of each specimen and wired in a wheatstone bridge arrangement with a dummy specimen to compensate for temperature.

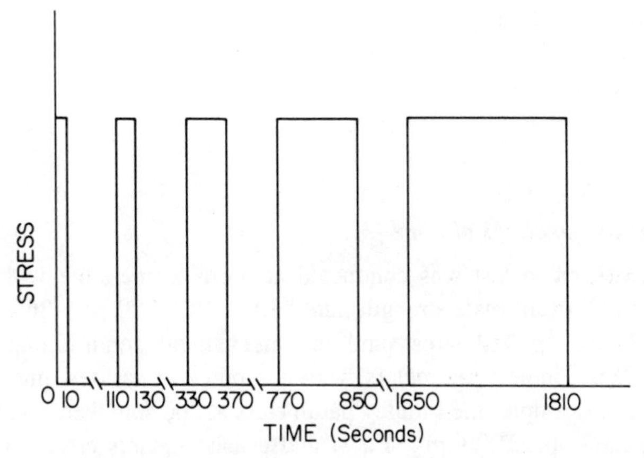

FIG. 2—*The mechanical conditioning procedure suggested by Jerina et al (Ref 9).*

Four stress levels were used for the creep tests: 11, 20, 30, and 36% of the static ultimate strength of SMC-R50 at room temperature. Prior to each creep test, a specimen was mechanically conditioned at a stress that was 50% of the ultimate strength at the test temperature.

Specimens were kept in a desiccator prior to testing. For room temperature tests, the specimens were placed in the oven which contained desiccant. Since all specimens had been dried in a desiccator for approximately two months prior to testing, it was assumed that the effect of moisture content was negligible.

Four temperature levels were used in this study: 23°C (73°F), 40°C (104°F), 70°C (158°F), and 100°C (212°F). Before performing elevated temperature creep tests, specimens were allowed to soak at the desired temperature for several hours. One specimen was used at a given temperature for all stress levels at that

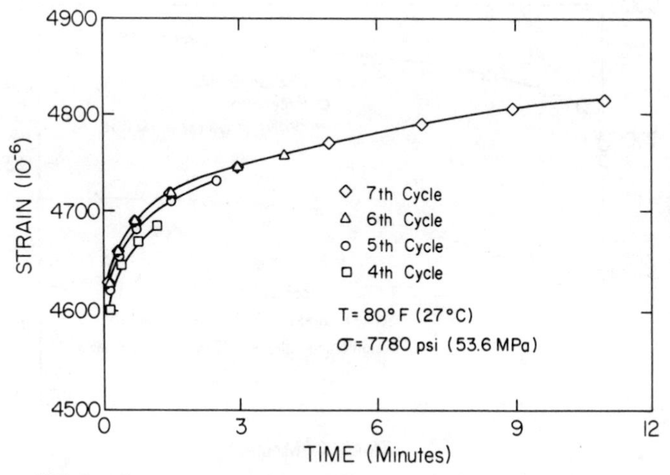

FIG. 3—*Creep response of SMC-R50 during mechanical conditioning.*

temperature. Thus, four specimens were used; replicate tests were not performed. Strain was allowed to recover before the next test was started. After recovery, a specimen was again mechanically conditioned before proceeding to the next test.

Results

Parameters as Functions of Time

A one-week creep test was conducted at room temperature with 44.9 MPa (6510 psi) (30% of ultimate strength) and 53.6 MPa (7780 psi) (36% of ultimate strength) as the applied stress and the mechanical conditioning stress, respectively. The Findley parameters were computed at various times during the creep test. For example, the Findley parameters at 200 min were found using the strain-time data up to 200 min, Eq 1, and a least-squares error computer program. Similarly, the strain-time data up to 1000 min were used to calculate the Findley parameters at 1000 min. It was found that the time exponent n varied with the time (Fig. 4) but approached an asymptotic value after approximately 8000 min. Along with the variation of n with time, the creep data is also presented in Fig. 4. (More will be said about this later.) The parameters ε_0 and m were also found to exhibit a time-dependent behavior as shown in Fig. 5.

Each of the parameters n, ε_0, and m approached asymptotic values after approximately 8000 min. It was found that the asymptotic values of ε_0 and m were attained in about 200 min if the asymptotic value of n was used in the computational procedure for calculating the best fit to experimental creep data (Fig. 6). In other words, using the asymptotic value of n, the strain-time data up to 200 min, Eq 1, and the least-squares error method, ε_0 and m reached their asymptotic values in 200 min (Fig. 6) instead of 8000 min (Fig. 5). Since n is independent of stress, only one longtime test is needed to determine the asymp-

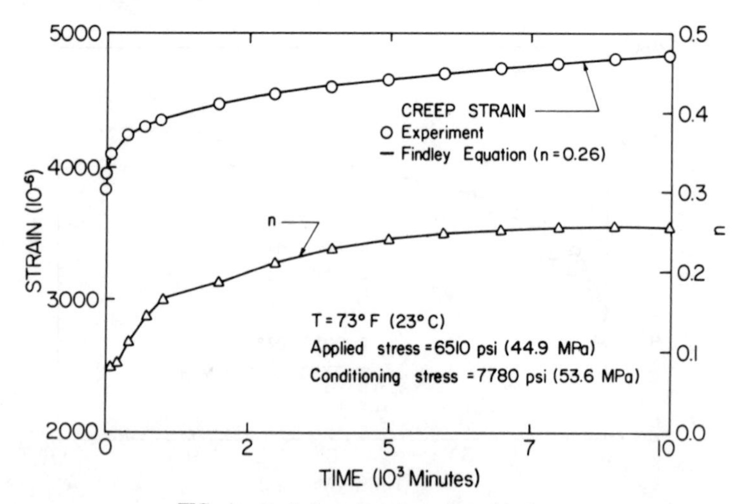

FIG. 4—*Variation of strain and* n *with time.*

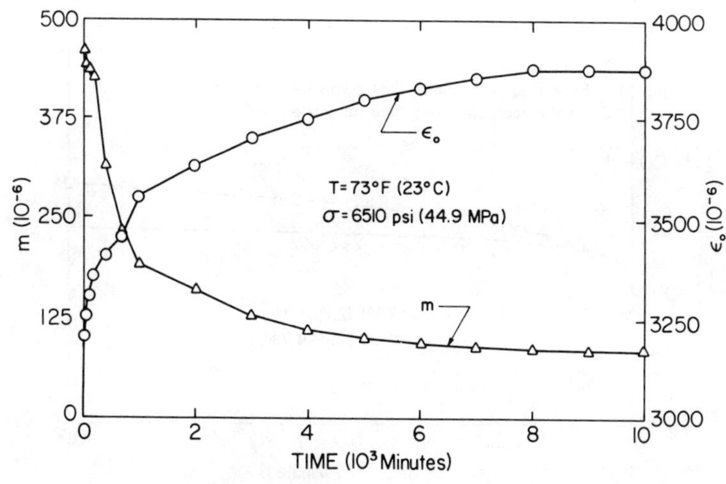

FIG. 5—*Variation of* ε_0 *and* m *with time.*

totic value of n. Using this value of n, ε_0 and m as functions of stress (at a given temperature) can be determined from creep tests on the order of 200 min duration. The predictive capability of the Findley equation using the asymptotic parameters is shown in Fig. 4.

The need to use asymptotic values of n, ε_0, and m is illustrated in Fig. 7. Clearly, evaluation of the parameters using short-time creep tests is inadequate.

Parameters as Functions of Stress and Temperature

Figure 8 shows the Findley parameters at four stress levels at room temperature. The specimen was conditioned at 53.6 MPa (7780 psi). The parameters

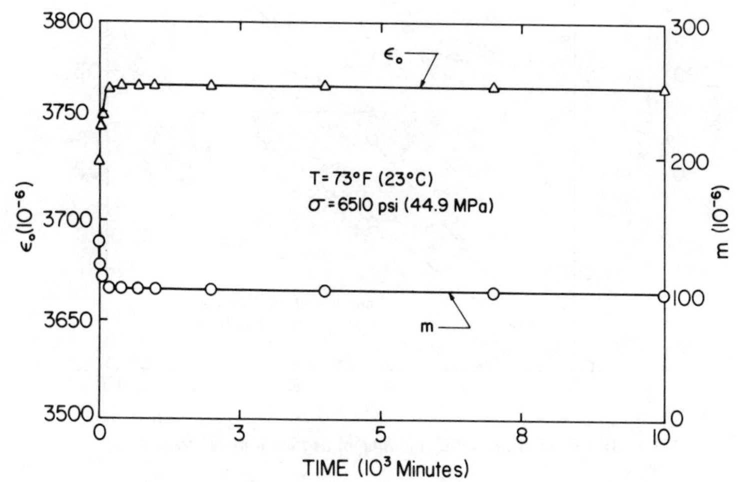

FIG. 6—*Variation of* ε_0 *and* m *with time using the asymptotic value of* n.

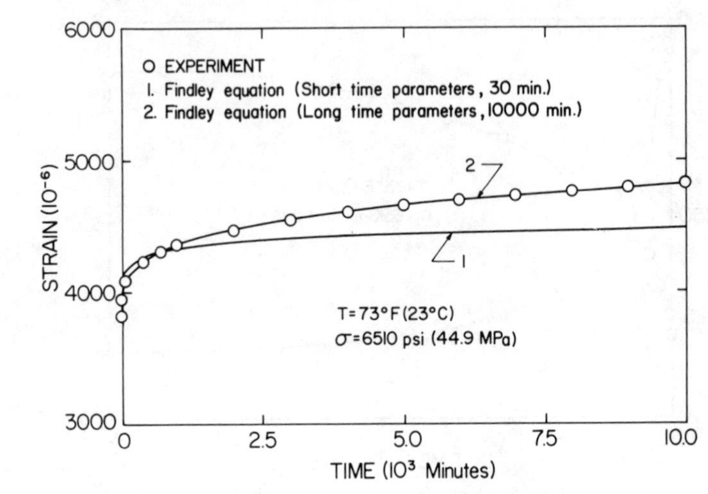

FIG. 7—*Prediction of creep response using short-time and longtime values of* n, ε_0, *and* m.

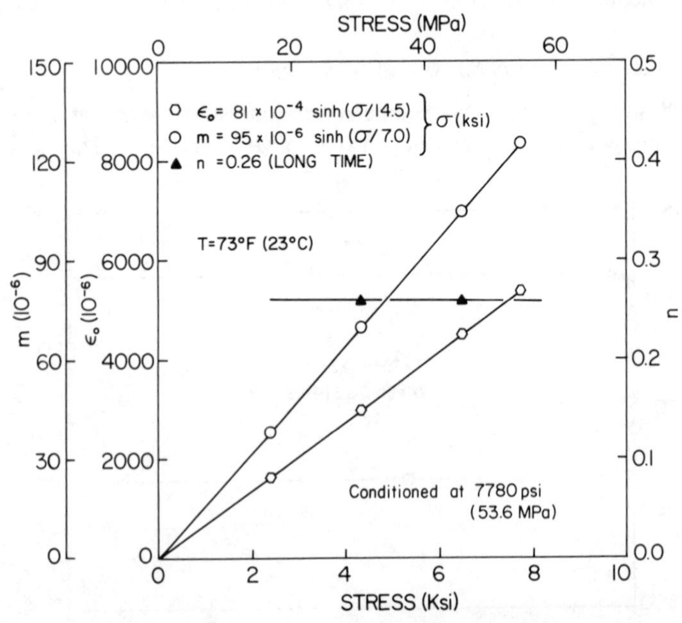

FIG. 8—*Parameter* n, ε_0, *and* m *as functions of stress.*

ε_0 and m can be described by hyperbolic sine functions of stress [*10*]. Creep tests of 200 min duration along with the asymptotic value of n were used to evaluate ε_0 and m.

For the elevated temperature tests, specimens were conditioned at 50% of the ultimate strength at each temperature. The longtime n value varied linearly with temperature as shown in Fig. 9. Similar results were found for glass-fabric laminates [*12*]. Using the longtime n values and 200 min creep data, the parameters ε_0 and m were calculated. Figures 10 and 11 show the hyperbolic sine fit of ε_0 and m, respectively.

The relationships between ε_0, m, and stress shown in Figs. 8, 10, and 11 are essentially linear. Thus, Eqs 2 and 3 could be replaced by linear relationships. Since the work reported herein is being extended to higher stress levels, the authors chose to use Eqs 2 and 3 instead of equally acceptable linear relationships.

Predictions Using the Findley Equation

Five longtime creep tests at elevated temperature were conducted. The duration of tests ranged from 10 000 to 30 000 min. Four of these tests were at 44.9 MPa (6510 psi); the results are shown in Fig. 12. Excellent agreement was observed between the predictions and the experimental data. Figure 13 shows the 30 000 min creep response at 37.4 MPa (5420 psi) where the temperature is 85°C (185°F). The maximum difference between experimental data and the Findley equation was less than 3%.

In addition to single-step loading, a multiple-step loading test was conducted. The loading consisted of a two-step loading at room temperature as shown in

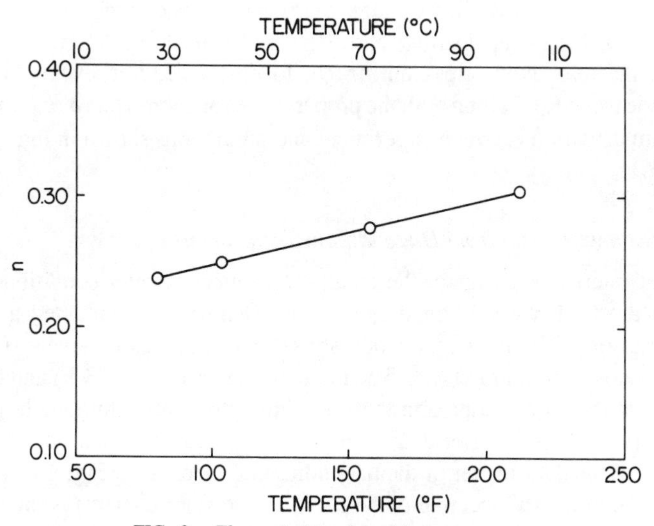

FIG. 9—*The variation of* n *with temperature.*

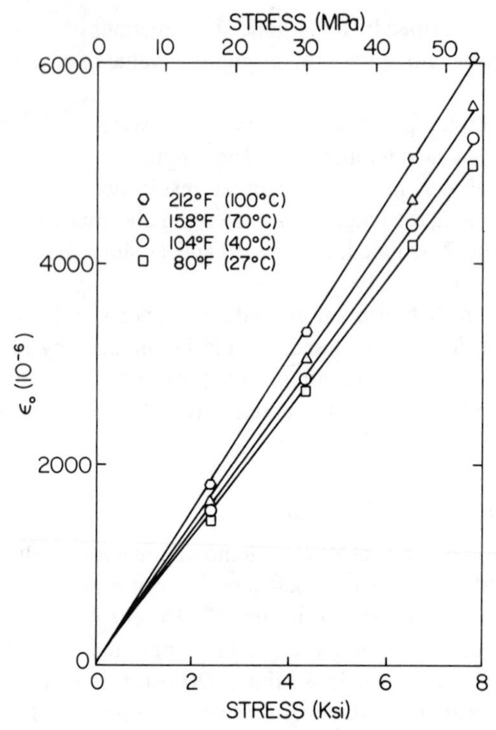

FIG. 10—ε_0 *as a hyperbolic sine function of stress.*

Fig. 14. The specimen was loaded at 44.9 MPa (6510 psi) for an hour; then the load was increased to 53.6 MPa (7780 psi) and held constant for another hour. Complete recovery was allowed after the release of the loads.

For the multiple-step loading case, the mechanical conditioning stress was equal to the maximum stress during the loading-unloading cycle. Predictions were made using Eq 7 along with the proper values of parameters n, ε_0, and m. The maximum deviation between experiment and predictions shown in Fig. 14 is less than 5%.

Comparison with Data from Unconditioned Specimens

It is of interest to compare the results from mechanically conditioned specimens to those of unconditioned specimens. Denton [*3*] conducted a series of creep tests on SMC-R50 at various stress and temperature levels. One such comparison is shown in Fig. 15. The temperature was 23°C (73°F) and the stress was 40% of the room temperature ultimate strength, both within the limits of the results reported herein. The data from a conditioned specimen lie at the upper boundary of the data scatter of the unconditioned specimens (Fig. 15). However, the exact condition of specimens [*3*] (such as moisture content) is unknown, so the comparison shown in Fig. 15 should only be considered qualitatively.

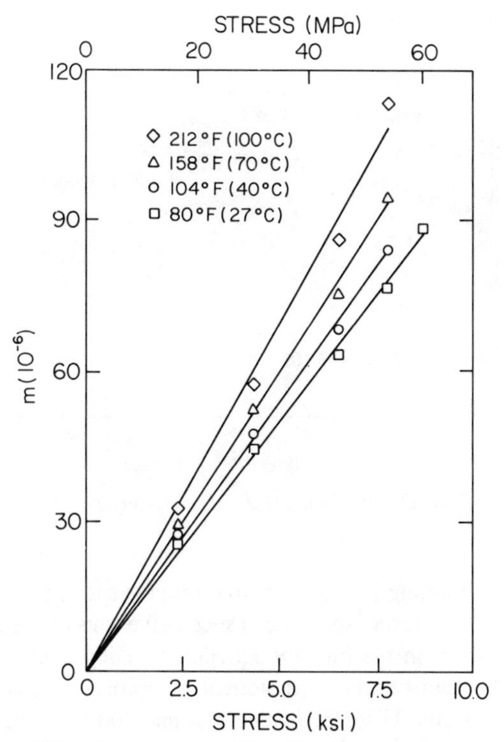

FIG. 11—m *as a hyperbolic sine function of stress.*

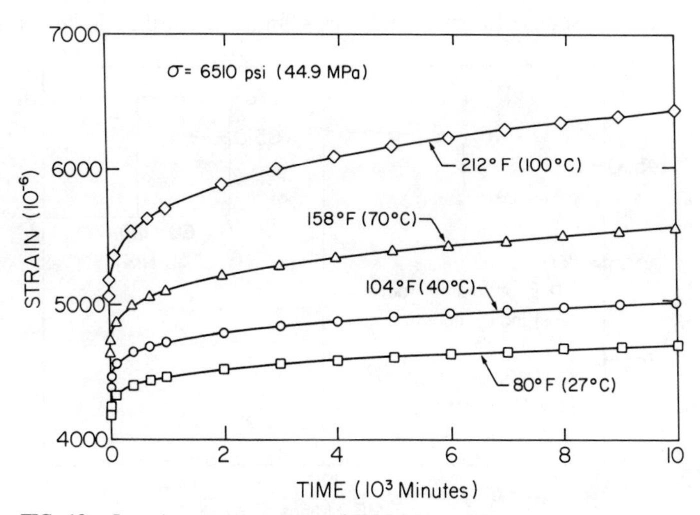

FIG. 12—*Longtime creep response of SMC-R50 at different temperature levels.*

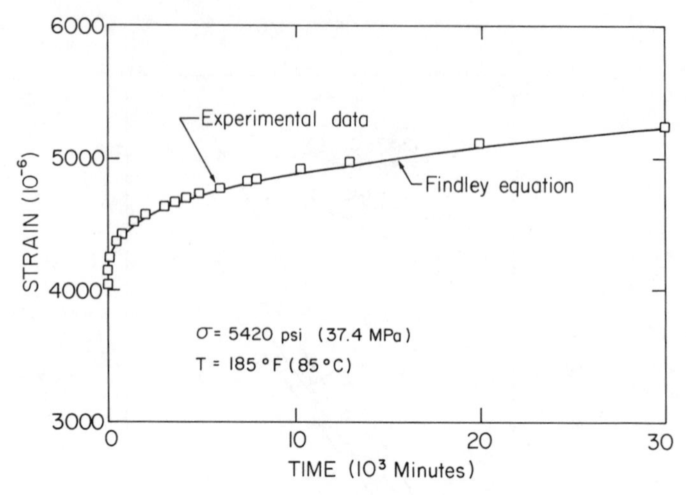

FIG. 13—*Prediction of longtime creep response.*

Conclusions

Mechanical conditioning is necessary to eliminate sudden jumps in strain and to give repeatable strain response between successive tests on the same specimen. The time exponent, n, in the Findley equation is independent of stress, varies linearly with temperature, and asymptotically approaches a constant value at approximately 8000 min. Using this value of n and 200 min creep data, parameters ε_0 and m are hyperbolic sine functions of stress and temperature. As previously stated, ε_0 and m could be also written as linear functions of stress for the data reported.

Within a limited range of stress and temperature, the Findley equation predicted the strain response for single-step tests. In addition, the Findley equation

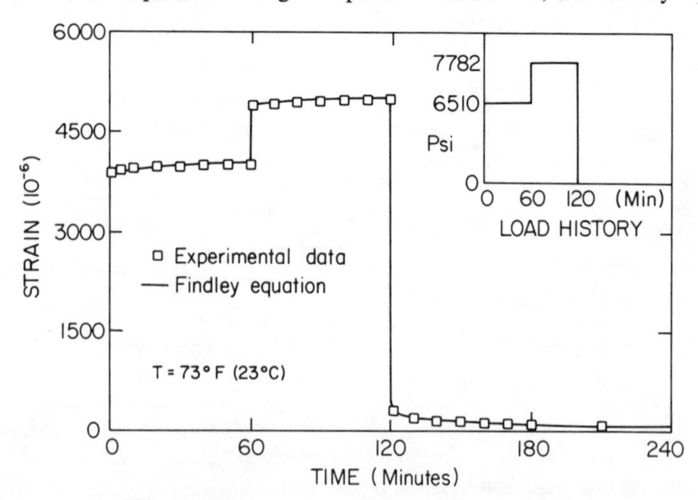

FIG. 14—*Creep response due to a two-step loading.*

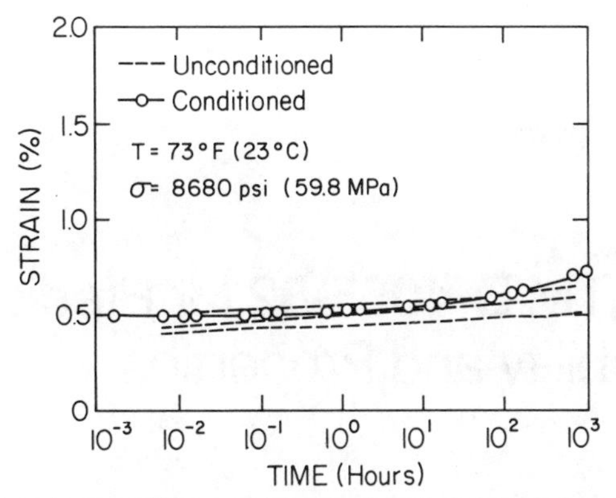

FIG. 15—*Comparison between the creep response of conditioned and unconditioned specimens.*

correctly predicted the strain response for a multi-step loading of increasing and decreasing loads. The strain response of conditioned specimens approximates that of unconditioned specimens.

References

[*1*] Heimbuch, R. A. and Sanders, B. A. in *Composite Materials in the Automotive Industry,* S. V. Kulkarni and R. B. Pipes, Eds., American Society of Mechanical Engineers, New York, Dec. 1978, pp. 111–139.

[*2*] Riegner, D. A. and Hsu, J. C., "Fatigue Considerations for FRP Composites," *Proceedings,* SAE Fatigue Conference, P-109, Society of Automotive Engineers, Detroit, MI, April 1982.

[*3*] Denton, D. L., "The Mechanical Properties of an SMC-R50 Composite," Composite Technology Update Publication, Owens-Corning Fiberglas Corp., Granville, OH, 1979.

[*4*] Riegner, D. A. and Sanders, B. A., "A Characterization Study of Automotive Continuous and Random Glass Fiber Composites," Technical Report GMMD 79-023, GM Manufacturing Development, General Motors Technical Center, Warren, MI, 1979.

[*5*] Christensen, R. M., *Theory of Viscoelasticity—An Introduction,* Academic Press, New York, 1971.

[*6*] Ferry, J. D., *Viscoelastic Properties of Polymers,* Wiley, New York, 1961.

[*7*] Cartner, J. S., Griffith, W. I., and Brinson, H. F. in *Composite Materials in the Automotive Industry,* S. V. Kulkarni and R. B. Pipes, Eds., American Society of Mechanical Engineers, New York, Dec. 1978, pp. 159–169.

[*8*] Schapery, R. A., *Journal of Polymer Engineering and Science,* Vol. 9, No. 4, 1969, pp. 295–310.

[*9*] Jerina, K. L., Schapery, R. A., Tung, R. W., and Sanders, B. A., "Viscoelastic Characterization of a Random Fiber Composite Material Employing Micro-Mechanics," MM3979-80-5, Texas A & M University, College Station, TX, 1980.

[*10*] Findley, W. N., Lai, J. S., and Onaran, K., *Creep and Relaxation of Nonlinear Viscoelastic Materials,* North-Holland Publishing Co., New York, 1976.

[*11*] Findley, W. N., Adams, C. H., and Worley, W. J. in *ASTM Proceedings,* Vol. 48, American Society for Testing and Materials, Philadelphia, 1948, pp. 1217–1239.

[*12*] Findley, W. N., Peithman, H. W., and Worley, W. J., "Influence of Temperature on Creep, Stress-Rupture, and Static Properties of Melamine-Resin and Silicone-Resin Glass-Fabric Laminates," National Advisory Committee for Aeronautics, Washington, DC, Technical Note 3414, 1955.

[*13*] Findley, W. N. and Lai, J. S., *Transactions of the Society of Rheology,* Vol. 11, No. 3, 1967, pp. 361–380.

Stephen Burke Driscoll[1]

Using ASTM D 4065-82 for Predicting Processability and Properties

REFERENCE: Driscoll, S. B., **"Using ASTM D 4065-82 for Predicting Processability and Properties,"** *High Modulus Fiber Composites in Ground Transportation and High Volume Applications, ASTM STP 873,* D. W. Wilson, Ed., American Society for Testing and Materials, Philadelphia, 1985, pp. 144–161.

ABSTRACT: Recently ASTM D 4065-82 (Practice for Determining and Reporting Dynamic Mechanical Properties of Plastics) was written to provide important rheological information on thermoplastics and thermosetting resins/reinforced plastics-composites. These procedures have been established for monitoring the subtleties of macromolecular behavior, the effects of processing, and in-service performance.

The significance of ASTM D 4065-82 is the generation of important rheological information that can be used to organize, monitor, and provide the necessary information for the quality control and assurance engineer.

Dynamic mechanical testing (DMT) of polymeric material systems provides important viscoelastic information for the materials, design, and process engineer. DMT quickly generates the critical information the materials engineer needs for formulating new composite systems, especially as raw material prices change. The design engineer will use DMT results for optimizing functional performance with greater reliability. The processing engineer will use DMT for assessing the effects of processing parameters on functional performance and determining the effect of changes in temperature on modulus, impact, and dimensional stability.

DMT provides critical information on the effects of the environment, such as moisture on long-term durability. Short-term dynamic mechanical characterizations can be used to establish "master curves" for predicting time-dependent behavior.

Representative examples of DMT as a function of frequency (dynamic oscillation), strain, temperature, and time are cited to illustrate how ASTM D 4065-82 is being used as an effective analytical procedure for ensuring quality and for predicting long-term functional performance. The final section of this paper details the use of instrumented impact testing [ASTM D 3763-79 (Test Method for High-Speed Puncture Properties of Rigid Plastics)] for generating additional information on functional performance.

KEY WORDS: plastics, rheology, rheological measurements, dynamic mechanical properties, curing behavior, plastic processability, thermosetting resins, composites, reinforced plastics

[1]Professor, Department of Plastics Engineering, University of Lowell, Lowell, MA 01854 and technical marketing consultant, Rheometrics, Inc., Piscataway, NJ 08854.

This paper is a review of a recently developed ASTM standard, Practice for Determining and Reporting Dynamic Mechanical Properties of Plastics (D 4065-82), and how the data generated in accordance with this standard can be used to organize, monitor, and provide important information for the quality control engineer [1].

ASTM D 4065-82 and its companion standard ASTM D 4092-82 (on Definitions and Description of Terms Relating to Dynamic Mechanical Measurements on Plastics) were developed by an ASTM Subcommittee D20.10.15 task group of several representatives of suppliers of resins, instrumentation systems, and convertors/fabricators/users of polymeric materials.

These two standards deal with the determination of important rheological properties of polymeric material systems. The standard practices outlined in ASTM D 4065-82 generate important information about the viscoelastic behavior of thermoplastics and thermosetting resins, about composites and about the influence of chemical additives.

Rheology is a scientific approach to solving manufacturing problems. It is both independent of other techniques and complementary in adding still another dimension — one more missing part — to the production quality puzzle. It is an effective bridge which spans the gap between material composition and functional performance and provides quality control information on incoming material which affects the manufacturer's outgoing product.

The literature is well documented to show how subtleties in polymer composition — structure/property relationships — can affect functional performance. Variations in molecular weight (MW), molecular weight distribution (MWD), and polymer branching have been detailed using rheological procedures. Major changes in end-use properties have been attributed to these differences. The popularity of DMT can be attributed to several factors, including the use of microprocessors to determine and measure more easily the sensitivity of molten polymer viscosity as well as the solid property behavior.

Of paramount importance to materials, process, and design engineers is the confidence in the data generated using dynamic mechanical measurements. The ASTM D 4065-82 standard provides the techniques to generate reliable, repeatable, and reproducible quality control information in just a few minutes. Sample preparation is fast, minimizing a source of testing error. Consequently, supplier and processor can interface without unreasonable concern for misinterpreting test information.

Dynamic mechanical measurements can be used on a wide range of polymeric materials, including soft, elastomeric compositions, hard and brittle plastics, or extremely rigid reinforced composites. Measurements in dynamic torsional shear and linear tension/compression as well as in three-point bending complement other rheological data generated using steady shear and extensional techniques. Thus, rheological quality control techniques are universal in scope and versatile in applicability. Dynamic mechanical analyses are based on forced oscillation of

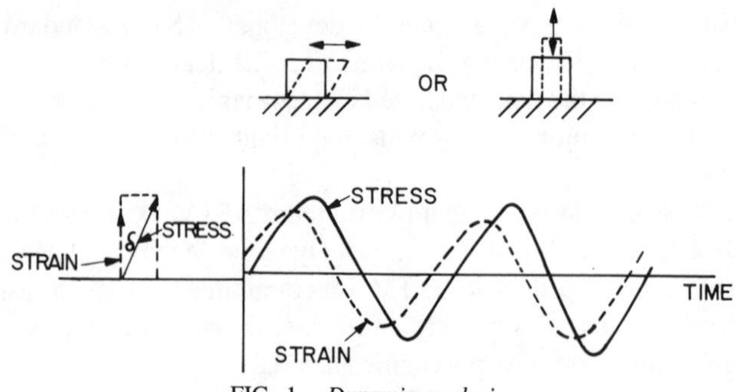

FIG. 1—*Dynamic analysis.*

a specimen and the measurement of the response in either a fixed or natural resonance mode (Fig. 1). Plastics are viscoelastic, that is, neither completely elastic nor viscous but a little of both and not always in the same proportion. A truly elastic material when deformed and released would return all the kinetic energy needed to deform it and regain its original shape. On the other hand, a purely viscous material when deformed and relaxed will "forget" its original configuration by dissipating the kinetic energy in the form of heat, flow, and other mechanisms. Consequently, since plastics are viscoelastic, exhibiting an in-phase elastic response and an out-of-phase loss component, it may be necessary to measure both attributes. The traditional measuring of the viscous component can be misleading since only part of the polymer's behavior is characterized. The elastic or storage behavior is especially critical since it may correlate to major manufacturing problems.

Dynamic mechanical measurements can be made on a wide range of physical geometries, using the traditional cone and plate and parallel disk configurations for polymer solutions and melts (Fig. 2) as well as rigid molded bars or rods. Additionally, the procedures in ASTM D 4065-82 accommodate polymer films and fibers (dynamic tension), elastomers and foams (dynamic compression), and three-point bend tests for characterizing high modulus composites or rigid but brittle materials (Fig. 3).

A wide range of specimen geometries is extremely useful since it allows a plastics material to be characterized for modulus and damping behavior over a

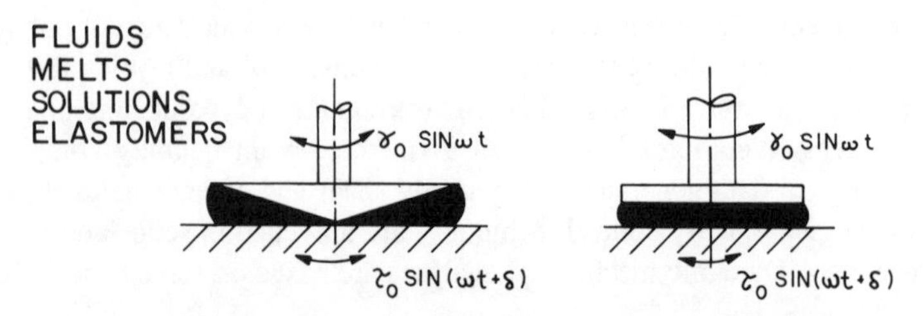

FIG. 2—*Dynamic analysis of cone and plate and parallel plate geometries.*

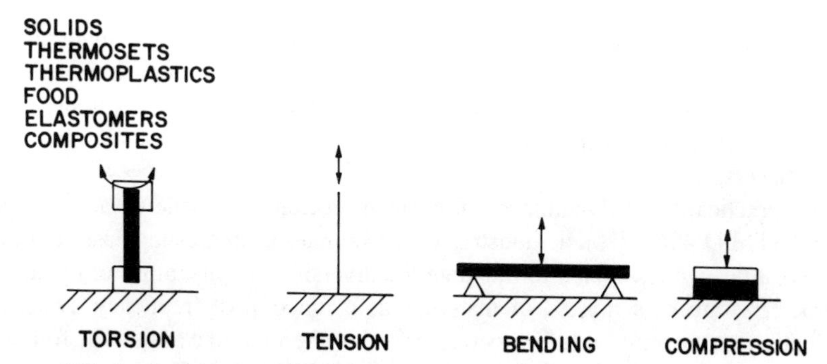

FIG. 3—*Dynamic analysis of other geometries for dynamic mechanical testing.*

broad temperature range; a simple and rapid changeover of test equipment is an important advantage of the typical analytical instrumentation used.

A typical representation of dynamic mechanical behavior over a wide temperature range is noted in Fig. 4. The modulus as a function of temperature scan is useful in generating important information about significant changes in material behavior, including (1) the glass transition (the temperature corresponding to the polymer's changing from a hard and brittle "glassy" to a soft, ductile, "leathery and rubbery" material) with the complementary tan delta peak (ratio of G'' to G') and (2) the secondary or beta transition which relates to impact strength and creep resistance.

The use of conventional cone and plate and parallel disks provides for immediate characterization of a molten polymer and generates important information used to control the processing of the material. This is especially significant since many polymers are very susceptible to excessive shear and thermal histories as well as to degradation due to moisture and oxidation. The torsional bar geometry is used to generate functional performance information, including modulus, impact strength, and creep resistance. Additionally, a plot of modulus versus tem-

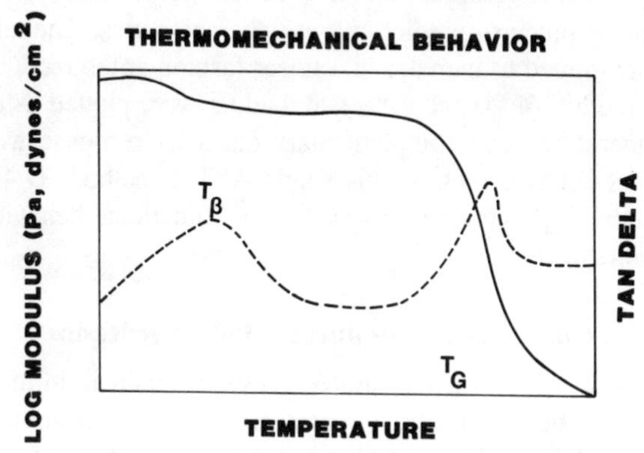

FIG. 4—*Thermomechanical properties as a function of temperature.*

perature behavior indicates the continuous service temperature of a material [at 10^{11} Pa (10^{10} dynes/cm^2)], and the impact strength of the composition can be directly related to the size, shape, and location of the beta peak transitions. This information is also useful in characterizing the uniformity of compounding and filler dispersions.

The practicality of dynamic rheological measurements made in accordance with ASTM D 4065-82 is demonstrated by examples of interesting case studies. These studies were selected to illustrate the diversified applicability of dynamic characterization of a myriad of materials as a function of frequency dynamic oscillation, strain amplitude, temperature, and isothermal time trials. The following comments underscore the significance of ASTM D 4065-82 as a simple means of characterizing the cure behavior of neat resins and prepregs. Only a few grams of material are needed for optimizing processing conditions based on research and development data.

Case Study One: Mineral-filled Polypropylene (PP)

A materials engineer wanted to use a chemical treatment [coupling agent (CA)] to improve the impact strength of a talc-filled thermoplastic resin. A secondary consideration was the effect of this surface treatment on the processability of the resin. A series of different types and concentrations of CAs was evaluated for complex viscosity and elastic modulus. The concentration of the CA was critical—too little or an excessive amount would seriously affect the rheological behavior.

The CA influences the complex viscosity, which relates directly to processability. A programmed strain sweep was also used to evaluate the uniformity of the filler in the base resin. This type of information can be used by toll-compounders for monitoring the quality of dispersed additives and the effects of using different types of compounding equipment, for example an intensive mixer versus twin-screw extruders.

The functional properties of the talc-filled PP were also evaluated using ASTM D 4065-82 protocols. A compression- or injection-molded test bar was evaluated over a temperature range to predict end-use characteristics. The temperature in Fig. 5 was programmed to increase in a linear fashion at the recommended 3 to 5°C/min. The moduli of 20 and 40% talc loading were plotted as a function of increasing temperature. The complementary tan delta response was noted and related to impact behavior. Thus, this single ASTM method, D 4065-82, was used to generate both processing information and functional behavior—all with only one instrument!

Case Study Two: Cure of Neat Unsaturated Polyester Resin

The curing behavior of an unsaturated polyester resin is influenced by the amount and type of the initiator system used as well as by the environment. The basic building blocks of the resin (the saturated and unsaturated acids or anhy-

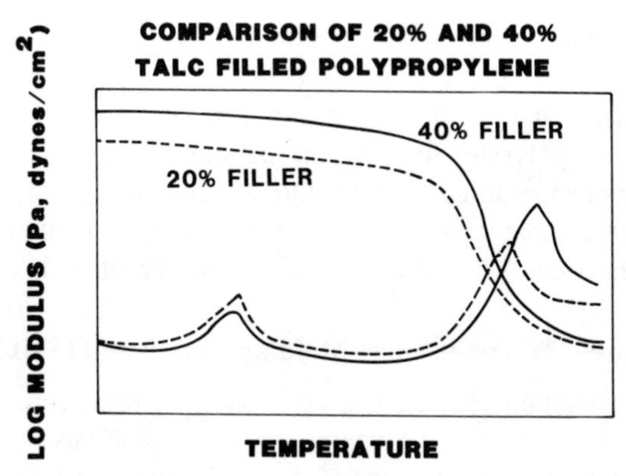

FIG. 5—*Modulus of talc-filled polypropylene as a function of temperature.*

drides, the polyol, and the reactive monomer) will influence the development of the three-dimensional polymer network and the functional properties of the cured polymer matrix. Using only a small amount of material (25 mm in diameter and about 3 mm thick), the engineer can command, through the interactive terminal, the temperature profile, frequency sweep, and strain amplitude to simulate actual processing conditions. This is, consequently, a very fast and inexpensive technique for optimizing formulations and for assuring quality control. Figure 6 illustrates the influence of the resin type and the amount of initiator on the cure behavior of the resin. The initial complex viscosity, n*, decreases as the material is heated to the isothermal equilibrium level of 120°C. The resin readily flows at

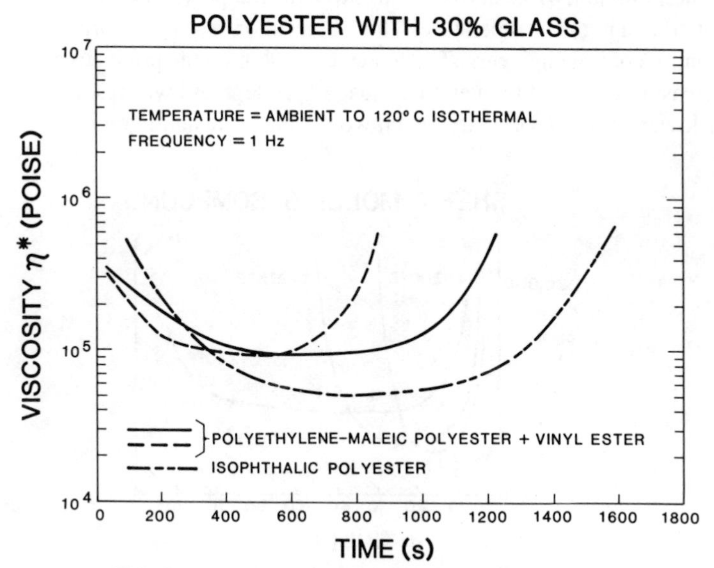

FIG. 6—*Cure of unsaturated polyester resin formulations.*

its minimum viscosity; this is followed by an increase in viscosity as the resin develops its structure. The rate of viscosity buildup is attributed to the resin matrix composition and the amount of initiator used.

The isophthalic polyester formulation exhibited a significantly lower equilibrium viscosity and required more time to undergo curing, while the vinyl ester formulation cured more quickly at an equivalent percent initiator and had a substantially faster cure at an increased concentration of initiator.

Case Study Three: Polyester Sheet Molding Compound (SMC)

After the formulation chemist has optimized the resin cure behavior (Case Two), he must then identify the interactions of the various chemicals used to modify the base resin. The influence of the reinforcing agent, the mineral fillers as extenders, and the coupling agents must be qualified. These additives will seriously alter the complex viscosity of the composite. The influence of temperature during cure must be documented to assist production quality control. Figure 7 illustrates the effect of temperature on the cure behavior of an unsaturated polyester resin which has been formulated for sheet molding. This sheet molding compound (SMC) was processed in matched-die tooling at three temperatures: 125, 140, and 150°C.

Two important trends have been identified in this study:

1. The complex viscosity of the SMC decreases, levels out, and increases rapidly with increasing mold temperature. This is understandable since the physical and chemical conversion of the SMC is quite temperature sensitive. The cooler the mold, the longer it will take to "kick off" the chemical reaction.

2. The minimum viscosity is much lower for the polymer composition which was cured at a higher temperature. This observation is very important to production quality control engineers. The tendency to shorten the production cycle time by increasing the mold temperature "just a few degrees" will result in a significantly lower complex viscosity which would lead to resin migration or weeping.

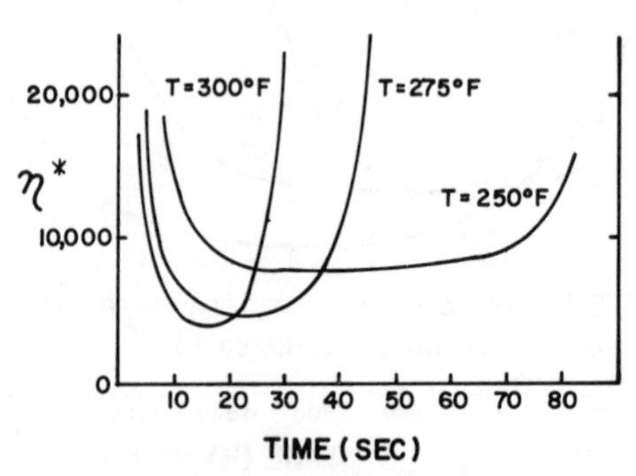

FIG. 7—*Effect of temperature on the cure behavior of unsaturated polyester sheet molding compound.*

Consequently, the cured SMC product might exhibit resin "rich" or "poor" areas, reduced physical properties, and loss of product quality.

The functional properties of these three materials could be easily determined using the solid test configuration accommodations in ASTM D 4065-82.

Case Study Four: Cure of Epoxy Resins as a Function of Hardener Type

ASTM D 4065-82 is not limited to only thermosetting polyester resins. The same concerns are valid for monitoring the cure behavior of bisphenol A-epichlorohydrin epoxy resins. Again, the selection of the type and concentration of the amine, acid, or anhydride curing agent or "hardener" is important in controlling viscosity before and during cure as well as with regard to the functional properties of the cure product such as a printed circuit board.

Figure 8 illustrates how the complex modulus, G* can be changed by varying the resin formulation. Akin to polyester cures, the hardener does influence the complex viscosity profile and the rheological properties of the functional composite.

Figure 9 illustrates how the ASTM D 4065-82 procedures provide the key information needed to formulate for specific manufacturing technologies. These data by Maximovich and Galeos [*2*] reiterate the observations of Tung and Dynes [*3*] in identifying the excellent correlation found between laboratory measurements and processing characteristics observed in the manufacturing shop. The significance of the intersection of G′ (the elastic or storage modulus) and G″ (the loss or viscous modulus) has been documented as a powerful analytical tool in identifying the gel time needed for streamlining the manufacturing development process.

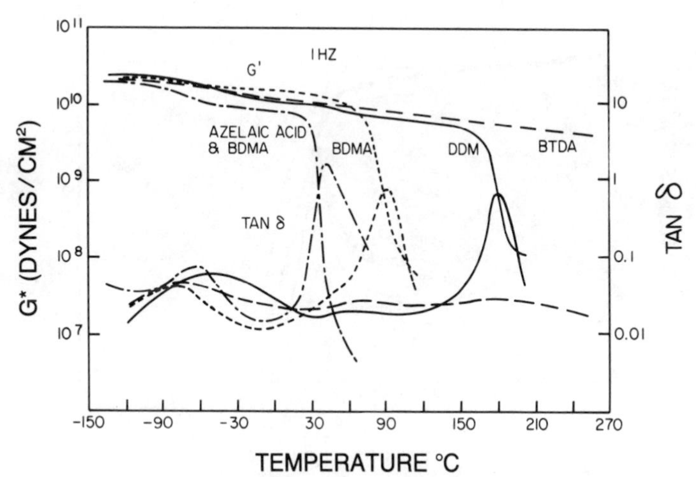

FIG. 8—*Effect of hardener on the cured behavior of epoxy resins.*

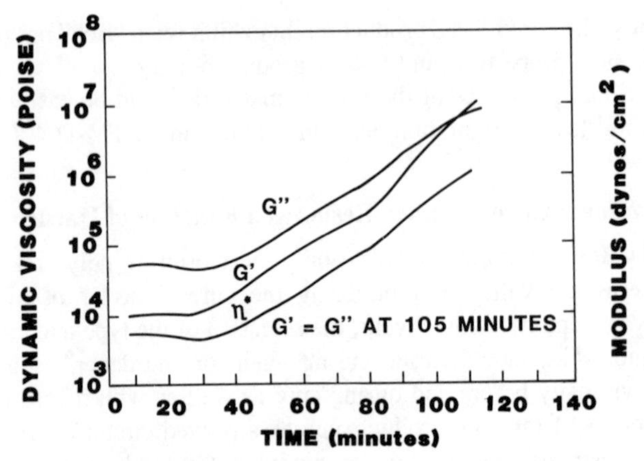

FIG. 9—*Cure profile of an epoxy resin: intersection of G' with G" to indicate gel point.*

Case Five: Base Resin Modifications

The commercialization of polymeric alloys or blends (PA/Bs) has not been limited only to thermoplastic systems. Figure 10, generated by ASTM D 4065-82 procedures, depicts the viscoelastic characteristics of a multi-component thermosetting resin system. The incorporation of either a soft phase, such as a rubber matrix, or a high-temperature component, such as polyimide, will alter the flow behavior, cure characteristics, and functional performance of an epoxy-based composite system. The low modulus rubbery phase contributes

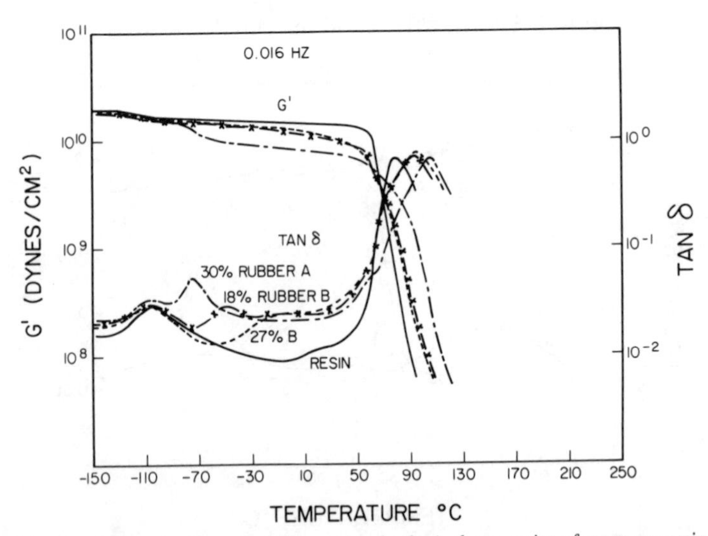

FIG. 10—*Effect of rubber modifier on the rheological properties of an epoxy resin.*

the expected improvement in impact strength (increased tan delta peak value at low temperatures). However, this enhancement of impact resistance is offset by a reduction in stiffness or rigidity when contrasted to the modulus of the base resin system at any specific temperature.

A polyimide component would contribute improved modulus at elevated temperatures. This would extend the functional ceiling temperature of the composite system, making it more attractive for engineering applications.

ASTM D 4065-82 quickly provides this important rheological information so that the materials engineer can formulate polymeric systems to meet exacting performance specifications. This can be achieved quickly and reliably using only small amounts of materials, which is indeed an economic advantage!

Case Study Six: Post Cure Treatment of a Thermosetting Composite

A processing engineer wanted to improve manufacturing productivity without compromising material performance. A series of polyimide laminates was cured and later oven-treated at 343°C for different time periods. The extent of oxidative cross-linking was determined by the enhancement of the structural rigidity, G', as a function of temperature. A mere 5-h thermal treatment significantly improved the modulus of the laminate. For example, the target modulus [10^{11} Pa (10^{10} dynes/cm^2)] temperature increased from 270°C up to an impressive 340°C and dramatically up to 380°C after only 5 and 12 h, respectively (Fig. 11).

Case Study Seven: Effect of Orientation of Prepreg on Functional Properties

In this case study, the rheological properties are plotted as a function of temperature (Fig. 12). These data illustrate the effect of reinforcement orientation

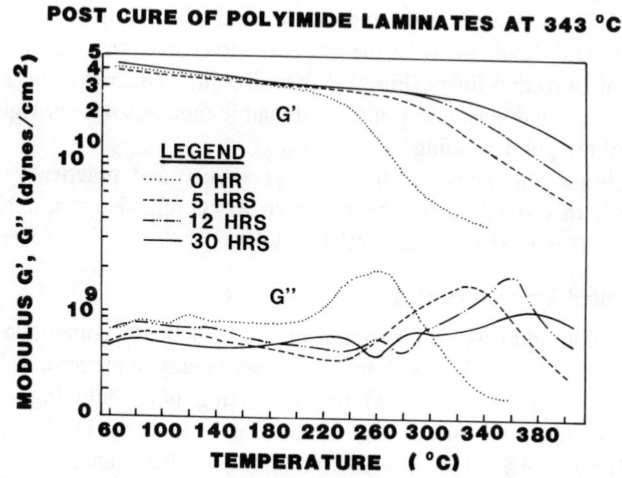

FIG. 11—*Effect of post-cure thermal treatment of a polyimide laminate.*

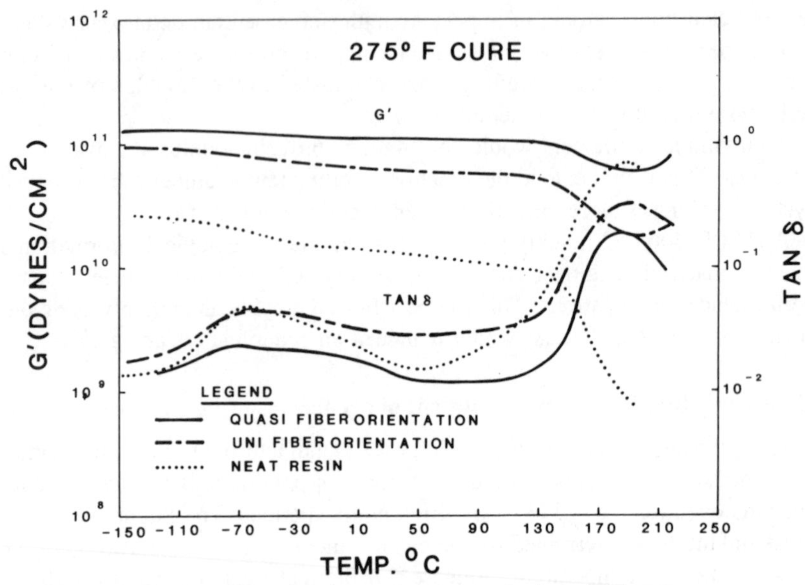

FIG. 12—*Effect of prepreg orientation on rheological properties.*

on the modulus of a cured composite system. Contrasted to the low modulus value of the neat resin, the uniaxially oriented prepreg does exhibit improved stiffness but its modulus is significantly lower than that of the random, quasi-oriented fiber construction. The tan delta response of these three candidates illustrates the effect of orientation on impact behavior. At low temperatures (−110 to −30°C) the neat resin and random construction exhibit greater tan delta values, which correlate to improved impact strength.

ASTM D 4065-82 embraces many different physical geometries and test modes to accommodate a wide range of materials. In addition to torsional modulus, ASTM D 4065-82 provides the procedure to characterize the dynamic tensile modulus of films and fibers (Fig. 13) as well as the dynamic compressive properties of foams and elastomers and the dynamic modulus of very rigid constructions via three-point bending.

These brief comments illustrate the versatility and practicality of ASTM D 4065-82; this standard has been shown to be reliable, repeatable, and reproducible. It is fast! It is efficient! It WORKS!!!

Instrumented Impact Testing

This section addresses the concern of the D20.10 Subcommittee for instrumented impact testing. This author has reviewed many of the principal objections and limitations of conventional impact testing [*4*], including ASTM Test Methods for Impact Resistance of Plastics and Electrical Insulating Materials (D 256-81) and ASTM Test Method for Impact Resistance of Rigid Plastic Sheeting or Parts by Means of a Tup (Falling Weight) (D 3029-82). Briefly, the

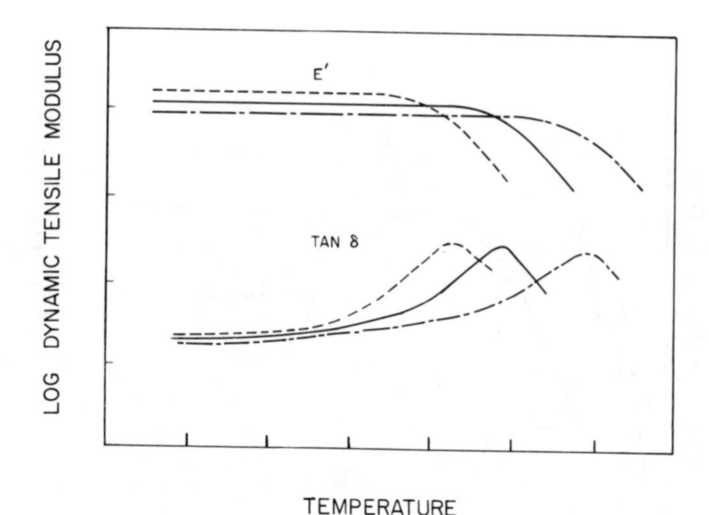

FIG. 13—*Elastic modulus in tension versus cure conditions.*

pendulum impact test, either Izod or Charpy, is only a uniaxial procedure which evaluates the singular directional properties of a laboratory specimen. The geometry of this coupon does not really represent the majority of real life situations, especially when impacted at a fixed speed of 3.46 m/s (0.1 mph or 11 ft/min). The effect of notching, the rate of notching, and the quality of the notch must be carefully determined for this simple geometry.

Dropped-weight testing does provide the capability to evaluate a specific area of a fabricated part. However, testing a "total" geometry is realistically limited to low drop heights. To impact at 13 m/s (30 000 in./min) would require a drop height of more than 9 m! Further complications of dropped-weight testing include the difficulty in identifying the criteria for failure as well as the multiplicity of samples required. The time and attendant expense results in a simple "mean failure height" and "mean failure energy" but no probability of failure! Figure 14 is a graphical representation of an instrumented driven dart impact test. While ASTM Test Method for High-Speed Puncture Properties of Rigid Plastics (D 3763-79) calls for a 1.25-cm (0.5-in.)-diameter tup impacting 7.5-cm (3-in.)-diameter supported samples, other geometries can be used. It is important to account for variations in specimen thickness, the shape of the impacting probe (flat, smooth radius, sharp point, etc.), and the ratio of the support ring to the impacting probe, which would simulate deflections ranging from shear punches to gentle deformations!

The rate sensitivity of the composite is also critical and should be identified over the complete impacting spectrum (Fig. 15).

The plot of the force [newtons (pounds of force)] versus deflection (millimetres, inches) provides substantially more important engineering information

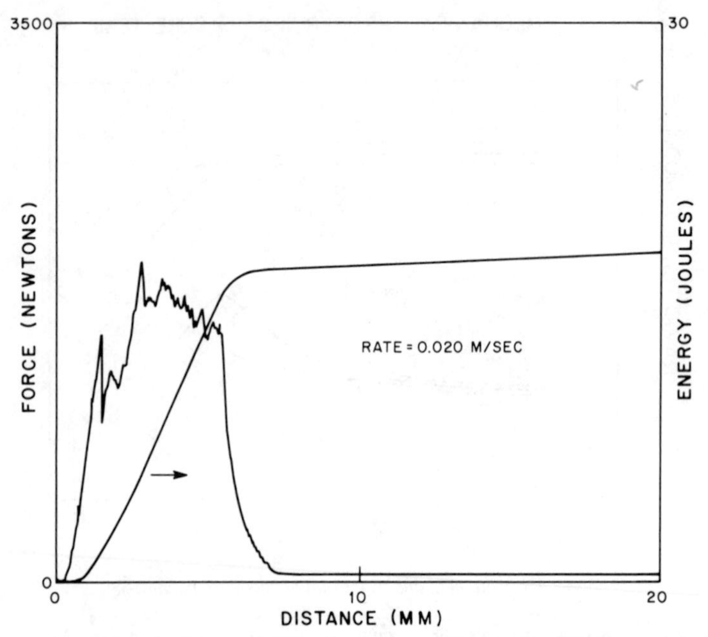

FIG. 14—*Instrumented impact testing: force versus deflection.*

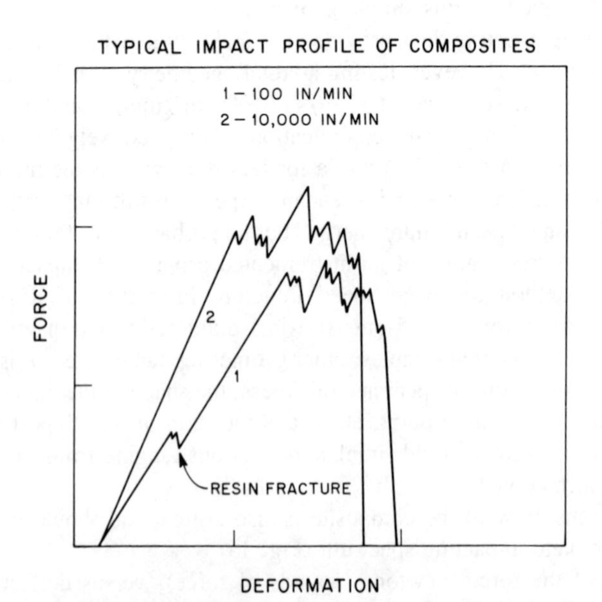

FIG. 15—*Instrumented impact testing: function of impacting speed.*

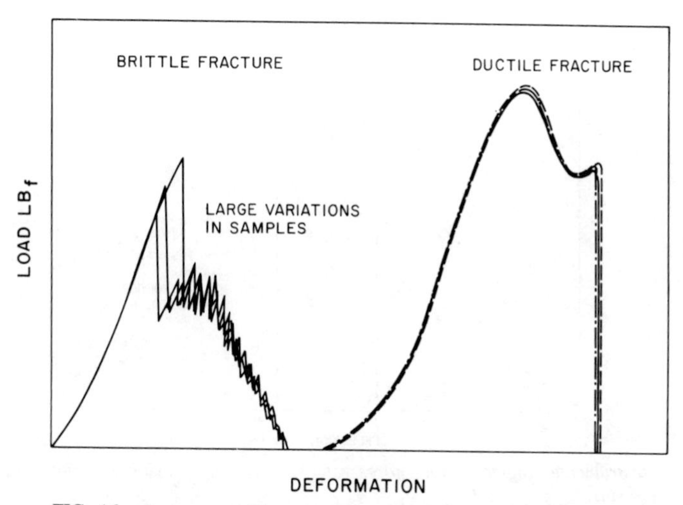

FIG. 16—*Instrumented impact testing: brittle versus ductile failure.*

useful to the design/materials/processing engineer. The automatic generation of energy [joules (inch-pounds of force)] at the energy deflection point complements the classification of the actual type of failure, that is, brittle versus ductile (Figs. 16–18).

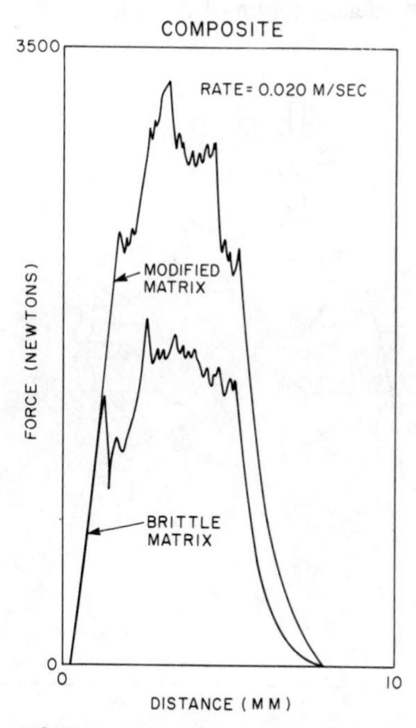

FIG. 17—*Instrumented impact testing: function of composition of resin formulation.*

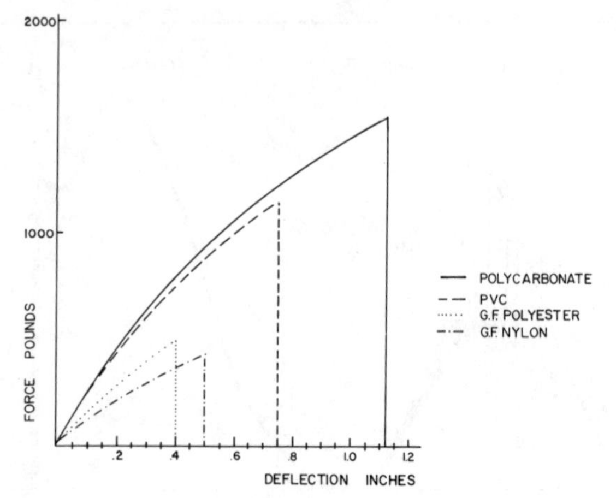

FIG. 18—*Instrumented impact testing: force versus deflection behavior of various engineering thermoplastic resins.*

The significance of ASTM D 3763-79 instrumentation is its sensitivity to a myriad range of materials, ranging from packaging materials to engineering thermoplastics (Fig. 18) to multi-ply composite constructions. Figure 19 shows the crack initiation, propagation, and probe-exit of an unsaturated polyester SMC. The individual peaks often correspond to the fracture development of the individual cracks and the delamination of the plies of the laminate.

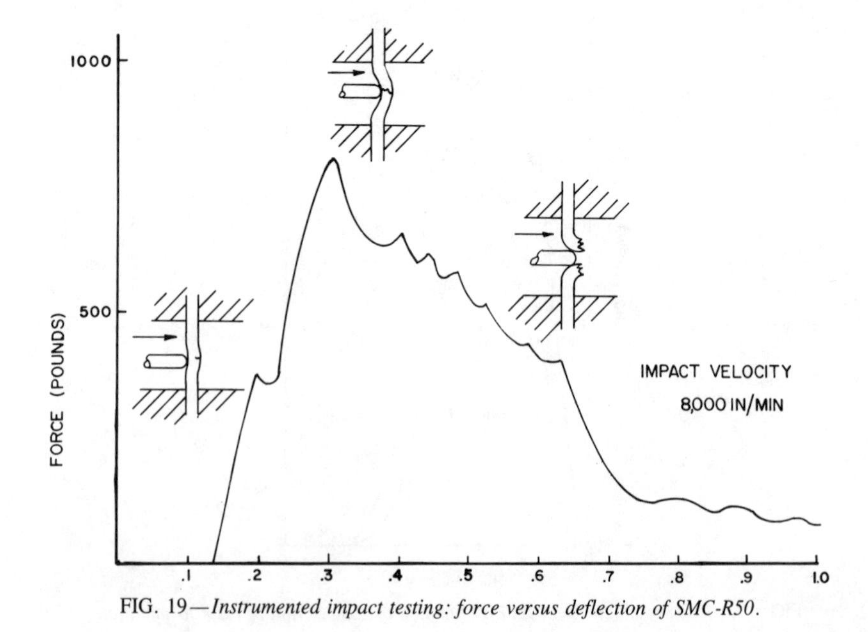

FIG. 19—*Instrumented impact testing: force versus deflection of SMC-R50.*

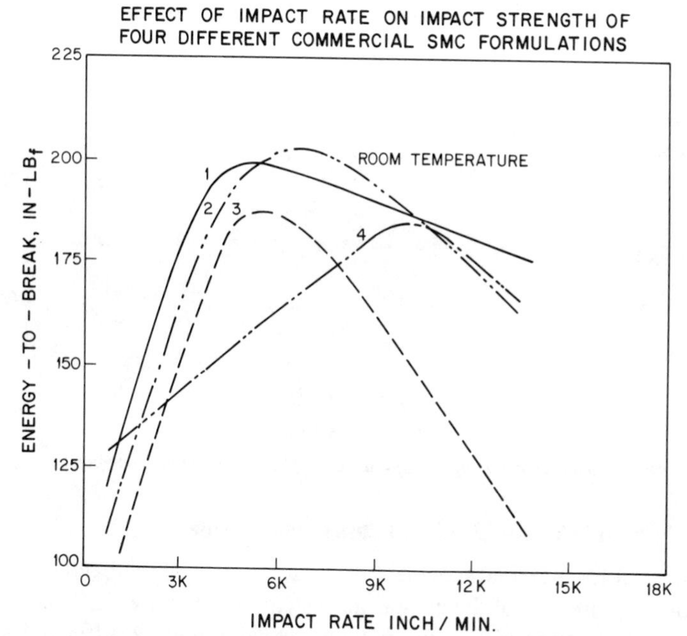

FIG. 20—*Instrumented impact testing: performance of SMC as a function of impacting speed.*

Case Study Eight: Rate Dependency of SMC

Figure 20 illustrates the rate sensitivity of four SMC formulations. The energy to break [joules (inch-pounds of force)] is plotted against the impacting rate [metres/second (inches/minute)]; all trials were conducted at ambient conditions.

Formulations 1, 2, and 3 exhibited approximately the same response at low impacting rates, below 2.16 m/s (5000 in./min). Compound 4 was significantly weaker. At 2.60 m/s (6000 in./min), Compounds 1 and 2 exhibited steady energy to break, while Compound 3 began to fail drastically and Compound 4 continued to improve linearly.

At 3.46 m/s (8000 in./min), the fixed impact speed of the pendulum test method, Compounds 3 and 4 exhibited the same energy to break. At 4.33 m/s (10 000 in./min), Compound 3 was a total failure while Compounds 1, 2, and 4 exhibited similar response.

These results emphasize why it is so important to avoid single data point tests. If the design engineer had selected a material based on an isolated impacting speed, it could have resulted in an expensive mistake. Since Compounds 3 and 4 exhibited the same energy to break at 3.4 m/s (8000 in./min), selecting Compound 3, for example, based on a lower compound cost, would have been a bad decision since Compound 4 did perform better at the faster impacting rates. Understandably, there can be no guarantee that ranking the energy to break these four SMCs would be consistent at other temperatures.

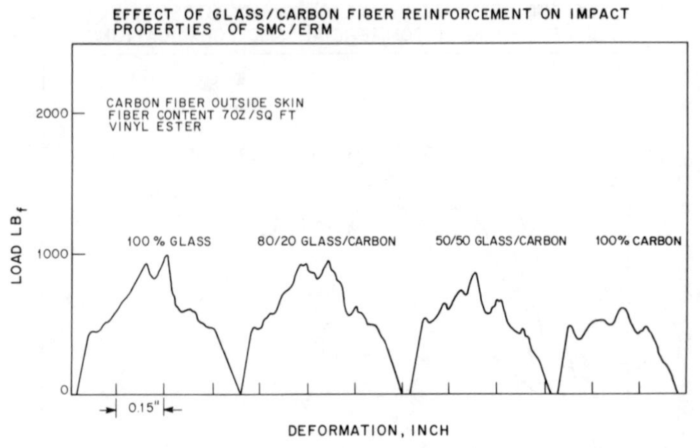

FIG. 21—*Instrumented impact testing: performance of SMC as a function of composite composition.*

Case Study Nine: Carbon/Fibrous Glass Constructions

The physical form and arrangement of the reinforcement will certainly affect the functional properties of the composite system. The type, weight and volume fraction, and orientation of the reinforcing phase will influence stiffness, rigidity, creep, and thermal resistance as well as impact energy behavior.

Figure 21 shows the significant differences in the force versus the deflection response of four composite systems containing varying ratios of fibrous glass and graphite fiber. Figure 22 depicts the effect of the arrangement of the reinforcing

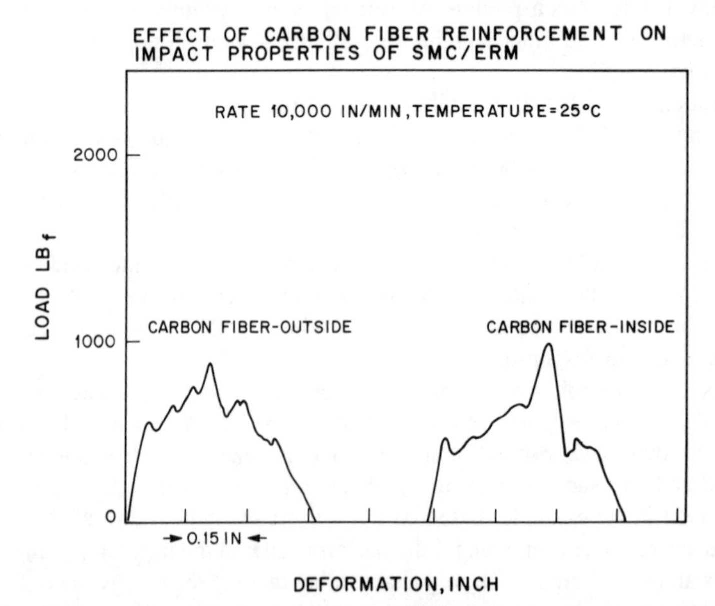

FIG. 22—*Instrumented impact testing: performance of SMC as a function of composite construction.*

agents. The composite based on graphite as an outer skin exhibited greater stiffness and lower ductility while the second example showed the contribution of the fibrous glass, which has relatively more elongation but lower rigidity at break. The total energy to break for these two samples is quite different.

Conclusions

These brief comments illustrate how ASTM D 4065-82 procedures are being used to characterize the cure behavior of neat resins and prepregs. The influence of reinforcing agents and other additives can be also easily identified, and the consequences of processing can be readily correlated with variations in functional properties, including modulus, creep resistance, and impact.

ASTM D 3763-79 is now being used to provide more meaningful information on impact behavior: how a part fails and its mechanism of failure, using a realistic geometry impacted at in-service rates. The advantages of instrumented impact testing include better accountability for brittle/ductile response as a function of rate, for temperature, and for the composition, fabrication, and thickness of the structural component.

References

[1] Driscoll, S. B. in *Quality Assurance of Polymeric Materials and Products, ASTM STP 846,* American Society for Testing and Materials, Philadelphia, 1983, pp. 83–102.

[2] Maximovich, M. G. and Galeos, R. M., "Rheological Characterization of Advanced Composite Prepreg Materials," Preprints, Society for the Advancement of Material and Process Engineering Conference, Vol. 28, April 1983, pp. 568–580.

[3] Tung, C. M. and Dynes, P. J., *Journal of Applied Polymer Science,* Vol. 27, 1982, pp. 569–574.

[4] Driscoll, S. B. in *Processing and Testing of Reaction Injection Molding Urethanes, ASTM STP 788,* American Society for Testing and Materials, Philadelphia, June 1981, pp. 86–93.

Summary

The technical papers presented in this book address some of the most difficult and important topics facing composites science today. These papers succeed in identifying key technical issues and providing insight and progress toward their solution. It is hoped that the numerous questions raised by the papers will serve to stimulate further research, for much remains to be done before a useful level of understanding is achieved.

One of the key differences between discontinuous high volume composites and the traditional aerospace continuous fiber systems is the control of properties through the control of fiber orientation state. The orientation of fibers in molded discontinuous fiber composites cannot be controlled well, and in most cases the orientation state changes unpredictably throughout the part. This, along with fiber distribution and overall material inhomogeneity, leads to large variability in material properties. The paper by Shirrell examines the relationship between the complex material microstructure in sheet molding compound and the variability in properties. For the R25 sheet molding compound (SMC) system, Shirrell's research showed that randomly occurring flaw sites in the material control strength. The failures observed occurred in the matrix, rarely breaking glass fibers. Additionally, a surface veil of individual fibers (that is, well dispersed and separated bundles) was manifested with a resultant increase in flexural strength. Specimens which upon microstructural analysis had the best fiber bundle separation and dispersion correspondingly had the lowest coefficients of variability. Shirrell also mentions the relationship of fiber orientation state to the measured elastic and strength properties but kept that parameter relatively constant in his studies. Denton and Munson-McGee expand upon the role of fiber orientation effects on the elastic properties of SMC and discuss a nondestructive technique for quantitatively measuring fiber orientation. The technique employs lead glass tracer fibers to "dope" the SMC so that X rays can be used to measure the in situ fiber orientation state. A high contrast photographic image of the X-radiograph is then used as a diffraction grating in a Fraunhofer diffractometer. The resulting diffraction patterns are then analyzed to quantitatively specify fiber orientation state. Since the fiber orientation state and distribution play such a key role in material property determination, the development of such a test method is extremely important.

Fiber orientation state and inhomogeneity also present some interesting challenges to the understanding of fatigue and fracture in discontinuous fiber composites. Mandell et al discuss an avoidance crack growth model, in which the crack extends by the coalescence of isolated failure zones, to predict fatigue crack growth and life trends in injection molded reinforced thermoplastics. For the materials investigated, submillimetre glass reinforced thermoplastics, a correlation was established between the model and experimental data. Fatigue crack growth was shown to follow a power law relationship, and the threshold fracture toughness values for crack growth in the composites were shown to be a higher percentage of K_{Ic} than that of unreinforced polymers or metals. Continuous fiber composites are also found in ground transportation applications, and the paper by Newaz discusses flexural fatigue behavior of E-glass/epoxy. Two distinct failure modes were identified corresponding to the deflection amplitude. High amplitude deflection produced matrix cracking, fiber breakage (tension side), and fiber buckling (compression side), resulting in rapid stiffness loss. Low amplitude deflection resulted in matrix cracking accompanied by fiber to matrix debonding on the tensile surface, causing gradual loss of stiffness. This research raises some interesting questions about the behavior of the material in a variable amplitude fatigue spectrum. The fatigue performance of composites is altered by the presence of environmental conditions. The combined effects of fatigue loading and exposure to water and *iso*octane on the fatigue life of SMC-R65 is discussed by Ngo et al. Water-soaked specimens showed about a tenfold decrease in fatigue life over dry specimens, while results for *iso*octane-conditioned specimens were inconclusive. Questions remain about the exact mechanism of property degradation caused by the liquid environment.

Bolted joint behavior in SMC-R50 is discussed in the paper by Wilson. While the strength of the joints was shown to vary with fastener diameter, torque, edge distance, and half spacing, scatter in the data makes it hard to quantitatively describe the relationship between variables. Throughout the test program, regardless of geometry, tension failure modes prevailed. Most failures were net tension, but for the large W/D specimens cleavage failures were observed. The large scatter of test data can be attributed to variable inhomogeneity in the material in combination with hole placement. The implications of this for design are far reaching and are further amplified in the paper by Gillespie and Pipes on the thermoelastic response of a cylindrically orthotropic disk. In this study on a transfer molded disk, the fiber orientation distribution was analyzed and the influence upon the thermoelastic properties described. It is shown that the molding process produces a disk with distinct layers of different fiber orientation state and that the layer thickness varies with radial position. A comparison between an analytical solution assuming homogeneous orthotropic properties and integration and finite element solutions accounting for the property variation reveals that the gross in-plane response is adequately predicted by all three methods, while there is an interlaminar free edge stress at the boundary which is accounted for only by analyses which handle the laminated structure of the disk and the radial variation in properties.

The combined effect of moisture and elevated temperature on the integrity of SMC-R65 is assessed in a paper by Hosangadi and Hahn. The paper focuses on the morphological characteristics of degradation by hygrothermal exposure, not on a quantitative assessment of property degradation. Their findings indicate that the infused water not only degrades properties by inducing mechanical strains but also by chemical mechanisms as evidenced by the examination of fluid found in blisters resulting from hygrothermal exposure.

Another important factor which influences final properties in both continuous and discontinuous composites is processing conditions. In order to optimize and develop consistant properties in high volume processes like filament winding and pultrusion, the flow and curing characteristics of the resin must be understood. Loos and Freeman present results on the influence of ply orientation on the flow characteristics of graphite epoxy laminates in their paper. They conclude that stacking sequence does not have a significant effect on the resin flow normal to the plane of the laminate, while results were inconclusive for effects on the flow of resin parallel to the plane of the laminate. The model developed by Loos and Springer applies as verified by comparison with experimental data.

If considered for structural applications, characterization of the time-dependent response of discontinuous fiber composites is necessary since the polymer matrix phase is highly viscoelastic. Yen et al characterize the viscoelastic response of an SMC-R50 material for different thermomechanical conditions. They found that the Findley equation adequately models the creep behavior and that the time exponent is independent of stress, varies linearly with temperature, and asymptotically approaches a constant value at approximately 8000 min. The results with lowest scatter were obtained by tests on conditioned specimens, a finding that has been previously reported for continuous fiber laminates by Shapery.

The final paper by Driscoll discusses the use of dynamic mechanical properties for predicting the processability and properties of plastics and composites (ASTM D 4065). The paper is in the form of an overview with case examples to explain the various test configurations and the material property parameters which can be measured using the test. Dynamic mechanical tests can be used to determine rheological parameters, viscoelastic properties, glass transition temperature, and temperature-dependent mechanical properties through shear, tension, compression, and flexural test configurations.

Composites are exciting materials because they offer the opportunity to simultaneously engineer the material and component, which if properly executed produces a degree of optimization in design not possible with isotropic materials. It is this potential which has spawned the use of composites in ground transportation and high volume consumer products. The research findings presented in this book have barely scratched the surface of knowledge which must and will be fostered in the coming years.

D. W. Wilson

Associate scientist and assistant director, University of Delaware, Center for Composite Materials, Newark, Delaware; symposium chairman and editor.

Index

G

H

I

L

M

N

P

R

S

T